Lecture Notes in Physics

Edited by H. Araki, Kyoto, J. Ehlers, München, K. Hepp, Zürich
R. Kippenhahn, München, D. Ruelle, Bures-sur-Yvette
H.A. Weidenmüller, Heidelberg, J. Wess, Karlsruhe and J. Zittartz, Köln
Managing Editor: W. Beiglböck

364

W0245891

Thomas T.S. Kuo
Eivind Osnes

Folded-Diagram Theory of the Effective Interaction in Nuclei, Atoms and Molecules

Springer-Verlag
Berlin Heidelberg GmbH

Authors

Thomas T. S. Kuo
Department of Physics, State University of New York at Stony Brook
Stony Brook, N.Y. 11794, USA

Eivind Osnes
Department of Physics, University of Oslo
P.O. Box 1048 Blindern, N-0316 Oslo 3, Norway

ISBN 978-3-662-13803-8 ISBN 978-3-540-46306-1 (eBook)
DOI 10.1007/978-3-540-46306-1

Originally published by Springer-Verlag Berlin Heidelberg New York in 1990
Softcover reprint of the hardcover 1st edition 1990

2153/3140-543210 – Printed on acid-free paper

PREFACE

In these notes we discuss in detail the folded diagrams which appear in the degenerate linked-diagram perturbation theory of the effective interaction in atomic nuclei. The emphasis is on the case with valence nucleons. Detailed diagram rules are derived and illustrated by simple examples. A general proof of the cancellation of disconnected diagrams is given, together with the derivation of many-body effective interactions. Although we are mainly concerned with the derivation of the nuclear shell-model effective interaction, we indicate that the method is rather general and can readily be extended to evaluate the nuclear optical model potential and the meson-exchange nucleon-nucleon interaction. Further, given the close similarities between nuclear structure and atomic and molecular structure calculations, we suggest that the method be applied to atomic and molecular problems. In fact, since the inter-particle interaction in atoms and molecules is the well-known Coulomb force and a large amount of precise experimental data exist for these systems, we point out that atoms and molecules may be a useful testing ground for the folded-diagram many-body theory.

These notes have developed from a number of lectures and seminars given at Stony Brook and Oslo and other places over an extended period of years. We are indebted to our colleagues at these institutions for useful discussions as well as stimulating scientific collaboration.

We are also grateful to our home institutions - the State University of New York at Stony Brook and the University of Oslo - for the support given to us over the years. Further, we acknowledge helpful support through grants from the US Department of Energy, the Norwegian Research Council for Science and Humanities (NAVF) and the Nordic Institute for Theoretical Atomic Physics (NORDITA).

Stony Brook
Oslo

1 June 1990

Thomas T.S. Kuo
Eivind Osnes

TABLE OF CONTENTS

1. Introduction

Although great resources have been directed towards relativistic heavy-ion collisions and intermediate-energy nuclear physics recently, the "old-fashioned" nuclear many-body problem† is still a central and fundamental problem in nuclear structure theory. The problem can be stated quite simply: Given the fundamental nucleon-nucleon interaction, can we calculate the properties of complex nuclei, such as for instance ^{16}O and ^{18}O?

In fact, during the last ten to fifteen years there has been considerable interest in the nuclear many-body problem, accompanied by several new developments. A detailed description of the many-fermion exp-S method was given by Kümmel, Lührmann and Zabolitzky[4]. A review of the Landau-Migdal theory for finite fermion systems, with applications to nuclei in the lead region, was presented by Speth, Werner and Wild[5]. In a series of three articles[6], a Green's function approach to the non-instantaneous, dynamic effective interaction between fermions in a many-body system was proposed and studied by Engelbrecht, Hahne and Heiss. Furthermore, the nuclear field-theory approach of Broglia and collaborators[7] has been very successful in correlating important nuclear properties.

However, none of the above theories are specifically designed for dealing with the effective interaction in the nuclear shell model, which for many years has served as an extremely useful and successful tool for describing nuclear properties and, very probably, will continue to do so in the near future. The purpose of the present notes is to give a detailed description of the folded-diagram method for the solution of a many-fermion problem. Within the framework of the non-relativistic Schrödinger equation, this method provides a systematic and convenient method for calculating nuclear properties from a given nucleon-nucleon potential. Although mainly used to study open-shell nuclei, this method is applicable to closed-shell nuclei

---

† Background material on this subject can be found in a number of textbooks, see e.g. refs.[1-3].

as well. A notable feature of this method is its close similarity to the empirical shell-model approach. Thus, the folded-diagram method appears to provide a microscopic foundation of the shell model and, in particular, of the derivation of the shell-model effective interaction starting from the fundamental nucleon-nucleon interaction.

All shell-model descriptions depend on the use of an effective residual interaction, since the eigenvalue problem is solved in a strongly truncated Hilbert space. Thus, it is necessary to use an effective interaction which differs from the true nucleon-nucleon interaction in that it includes the contributions from the excluded configurations. In early shell-model calculations[8)] simple phenomenological potentials were used for the effective interaction. All these potentials contained parameters which were determined so as to optimize the agreement with the experimental energy spectra. Very interesting developments were then made by Talmi and collaborators[9)], who were able to determine the matrix elements of the effective interaction in simple configurations directly from the experimental spectra. Both these approaches were highly successful in correlating and predicting nuclear data.

The early successes of the shell model prompted extensive efforts[†] to understand the physical origin of the shell model and, in particular, to calculate nuclear properties from "first principles", i.e. the free nucleon-nucleon interaction and many-body theory. Among the early calculations along these lines (to be referred to as realistic calculations) those of Kuo and Brown[13,14)], which were based on low-order perturbation theory, were especially encouraging. Since then, numerous attempts have been made to explain why low-order perturbation theory seems to work so well. These developments[††] included both calculation of individual higher order diagrams and summation of infinite subsets of diagrams in the perturbation expansion of the effective interaction. However, one has not yet been able to establish

---

† These are reviewed in refs.[2,10-12)].

†† See e.g. refs.[10-12)].

the convergence properties of this expansion[15-21] and the problem is still being investigated.

Most of the realistic calculations of nuclear properties are based on the folded-diagram expansion of the effective interaction. As starting point for the latter we may consider the Goldstone expansion, which just gives the one-dimensional effective interaction in non-degenerate systems. The Goldstone expansion has served as the basis of the Bethe-Brueckner-Goldstone theory of nuclear matter and is also applicable to the calculation of ground-state binding energies of closed-shell nuclei. For the calculation of open-shell nuclei (and excited states of closed-shell nuclei) we need, however, a degenerate linked-diagram perturbation theory. This requires the introduction of folded diagrams, which will be discussed in detail in these notes. We shall give an elementary and systematic treatment of folded diagrams and carefully investigate some rather subtle points about such diagrams. Special emphasis will be given to the case with valence nucleons, and detailed diagram rules will be formulated.

Folded diagrams may be introduced in several ways. Time-dependent perturbation theory was used by Morita[22], Oberlechner et al.[23], Johnson and Baranger[24] and Kuo et al.[25]. Brandow[26] started from a time-independent energy-dependent Brillouin-Wigner perturbation expansion, while Lindgren[27] showed that the energy-independent Rayleigh-Schrödinger perturbation expansion can be used. The formulation of Kuo, Lee and Ratcliff[25] (KLR) appears to have several advantages in being rather transparent in structure and convenient for actual calculations. This will become clear later. Thus, we shall follow rather closely the approach of KLR. It may be pointed out that a detailed description of this approach has not yet been given.† Thus, a partial purpose of the present work is to provide a comprehensive and relatively complete development of the KLR approach.

---

† There have been several papers[11,28,29] discussing the KLR approach and its extensions. But because of the nature and length of these papers, detailed derivations were not given.

A desirable feature of the KLR approach is its applicability to a wide range of physical problems. Although in these notes we shall mainly be concerned with the derivation of the nuclear shell-model effective interaction, the method can readily be extended to the study of the nuclear optical model potential[30)] and the free nucleon-nucleon potential[31)]. These subjects will be briefly discussed after we have discussed the fundamentals of the folded-diagram method. The method can also be applied to atomic and molecular structure calculations, as has been done by Yamamoto and Saika[32)] and others. Such applications will be discussed at the end.

As there is close similarity between the folded-diagram effective interaction and the empirical shell-model effective interaction mentioned above, we shall first briefly outline the shell-model approach to ^{18}O. Suppose we wish to calculate the properties of the low-lying states in ^{18}O. Then, we must solve the Schrödinger equation

$$H\Psi_\lambda = E_\lambda \Psi_\lambda, \tag{1}$$

where the Hamiltonian H can be written in the form

$$H = T + V = (T + U) + (V - U) \equiv H_o + H_1. \tag{1.1}$$

Here, T is the kinetic energy and V the nucleon-nucleon interaction. We take the eigenfunctions Φ_i of H_o as a basis for the expansion of the eigenfunctions Ψ_λ of H. Thus, it is customary to choose an auxiliary one-body potential U of convenient mathematical form, e.g. the harmonic oscillator potential

$$U = \sum_{i=1}^{A} \tfrac{1}{2} m\omega^2 r_i^2 . \tag{1.2}$$

The spectrum of H_o appropriate to ^{18}O is shown in fig. 1. For low-lying

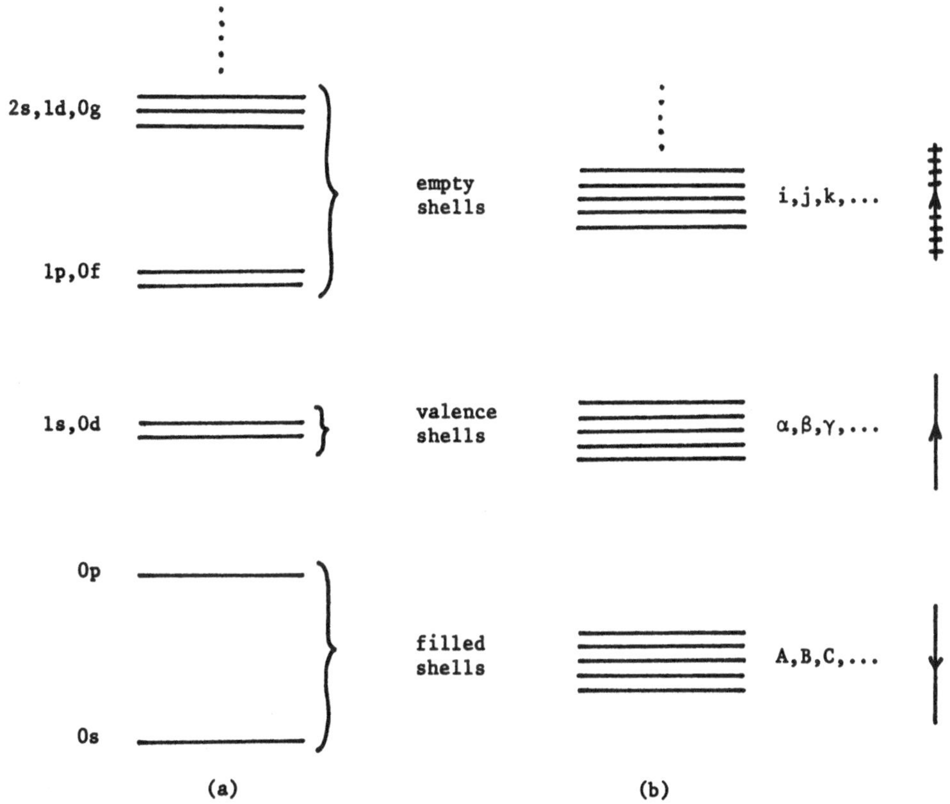

Fig. 1. Classification of shell-model single-particle orbits for the calculation of low-energy spectra of (a) ^{18}O and (b) an arbitrary nucleus near closed shells. Single-particle orbits (lines) belonging to the valence shells are called active orbits (lines) and single-particle orbits (lines) belonging to the empty and filled shells are called passive orbits (lines).

states of ^{18}O we expect the wave functions to be dominated by components with a closed ^{16}O core (i.e. the 0s and 0p orbits are filled) and two neutrons in the valence orbits 1s and 0d. Thus, we choose a model space which is spanned by the vectors

$$|\Phi_i\rangle = \sum_{\substack{\alpha>\beta \\ \in \text{ valence shells}}} C^{(i)}_{\alpha\beta}\, a^\dagger_\alpha a^\dagger_\beta |c\rangle , \qquad i = 1,2,\dots,d; \qquad (2)$$

where $|c\rangle$ is the unperturbed ^{16}O core obtained by completely filling the 0s and 0p orbits

$$|c\rangle = \prod_{A \in \text{filled shells}} a_A^\dagger |0\rangle . \tag{2.1}$$

To be more specific, the model space will be three-dimensional for the $J^\pi = 0^+$ states of ^{18}O, the three basis vectors being

$$|\Phi_1\rangle = \frac{1}{\sqrt{2}} [a^\dagger(0d_{5/2})a^\dagger(0d_{5/2})]^{J=0}|c\rangle ,$$

$$|\Phi_2\rangle = \frac{1}{\sqrt{2}} [a^\dagger(1s_{1/2})a^\dagger(1s_{1/2})]^{J=0}|c\rangle ,$$

$$|\Phi_3\rangle = \frac{1}{\sqrt{2}} [a^\dagger(0d_{3/2})a^\dagger(0d_{3/2})]^{J=0}|c\rangle . \tag{2.2}$$

Here, the square brackets denote coupling to good angular momentum

$$[a^\dagger(j_a)a^\dagger(j_b)]^{JM} = \sum_{m_a, m_b} \langle j_a m_a j_b m_b | JM\rangle \, a^\dagger(j_a m_a) a^\dagger(j_b m_b) . \tag{2.3}$$

If $j_a \equiv (n_a \ell_a j_a) = j_b$, an extra factor $1/\sqrt{2}$ is needed for normalization.

It is convenient to define projection operators P and $Q = 1-P$ which project from the complete Hilbert space onto the model space and its complementary space (excluded space), respectively. Then, P can be expressed in terms of the vectors (2) as

$$P = \sum_{i=1}^{d} |\Phi_i\rangle\langle\Phi_i| . \tag{3}$$

By means of this operator we can formally reduce the original eigenvalue problem (1) to the model-space eigenvalue problem

$$PH_{eff}P\Psi_\lambda = E_\lambda P\Psi_\lambda . \tag{4}$$

The main purpose of this work is to calculate H_{eff} from the original H. In fact, eq. (4) is very similar to the shell-model eigenvalue equation, except that in the shell model one calculates binding energies with respect to the core. Thus, we expect to have a model-space eigenvalue problem of the form

$$PH'_{eff}P\Psi_\lambda = (E_\lambda - E_C)P\Psi_\lambda , \tag{4.1}$$

where E_C is the true energy of the core; i.e. in the present case, the true ground-state energy of ^{16}O. The effective Hamiltonian is then

$$H'_{eff} = H'_o + v_{eff} , \tag{4.2}$$

where

$$H'_o = \sum_\alpha \varepsilon'_\alpha a^\dagger_\alpha a_\alpha \tag{4.3}$$

is the effective one-body Hamiltonian measuring the valence single-particle binding energies ε'_α with respect to the core. Empirically, ε'_α is taken as the binding energy difference between the state α in the appropriate nucleus with one nucleon in addition to closed shells and the ground state of the corresponding closed-shell nucleus. In the present example one would then take the differences between the binding energies of the lowest $5/2^+$, $1/2^+$ and $3/2^+$ states in ^{17}O and the ^{16}O ground state. The term v_{eff} in eq. (4.2) is the effective interaction between the valence nucleons, which is from two- to n-body character, n being the number of valence nucleons considered. In the ^{18}O example, v_{eff} is clearly the effective interaction between the two valence neutrons confined to the (1s0d) shell. In shell-model calculations v_{eff} is generally taken as a pure <u>two-body</u> interaction, even if there are more than two valence nucleons present. The matrix elements of v_{eff} may then be determined empirically so as to produce good agreement between calculated and measured energy differences $E_\lambda - E_C$. This method has been very successful in reproducing experimental energy levels and has thrown light on salient features of the effective two-nucleon interaction[9-12].

As pointed out above, we shall use degenerate time-dependent perturbation theory to derive a linked-diagram expansion for H_{eff} in eq. (4) for a general P-space. (Needless to say, if P is one-dimensional, this H_{eff} is identical to the H_{eff} given by the Goldstone expansion.) Then, in order to obtain an effective Hamiltonian appropriate to shell-model calculations, we shall reduce eq. (4) to eq. (4.1), so that the final effective Hamiltonian actually becomes H'_{eff}.

Before going on further, we may point out to our readers that we shall be dealing with a large number of diagrams, many being illustrative examples, in this paper. But this does not at all mean that the folded-diagram method is complicated. As a single diagram may be more worth than many words, we have included several diagrams just for the convenience of explaining things. The application of the folded-diagram theory to actual nuclear structure calculations is indeed very simple (see sect. 7). In fact, it is not necessary to evaluate any individual folded diagrams in most applications; we only need to calculate the non-folded diagrams and their energy derivatives.

2. Folded diagrams

In the present discussion we shall use the time-evolution operator in the so-called complex-time limit[1]. (Equivalently, we may use the evolution operator of the Bloch approach[33]. But for reasons which will soon become clear, we choose not to use the adiabatic[34] time-evolution operator.) In the complex-time limit we can write the perturbation expansion of the time-evolution operator $U(0,-\infty)$ in the interaction representation as

$$U(0,-\infty) = \lim_{t' \to -\infty(\varepsilon)} \sum_{n=0}^{\infty} \left(\frac{-i}{\hbar}\right)^n \int_{t'}^{0} dt_1 \int_{t'}^{t_1} dt_2 \cdots \int_{t'}^{t_{n-1}} dt_n \, H_1(t_1)H_1(t_2)\ldots H_1(t_n) \,, \tag{5}$$

or, with the introduction of the time-ordering operator T, as

$$U(0,-\infty) = \lim_{t' \to -\infty(\varepsilon)} \sum_{n=0}^{\infty} \frac{1}{n!} \left(\frac{-i}{\hbar}\right)^n \int_{t'}^{0} dt_1 \int_{t'}^{0} dt_2 \cdots \int_{t'}^{0} dt_n \, T[H_1(t_1)H_1(t_2)\ldots H_1(t_n)], \tag{5.1}$$

where $\lim_{t' \to -\infty(\varepsilon)} \equiv \lim_{\varepsilon \to 0+} \lim_{t' \to -\infty(1-i\varepsilon)}$.

Consider the wave function

$$U(0,-\infty)|\Phi_i\rangle = U(0,-\infty)\, a_\alpha^\dagger a_\beta^\dagger |c\rangle \to \tag{6}$$

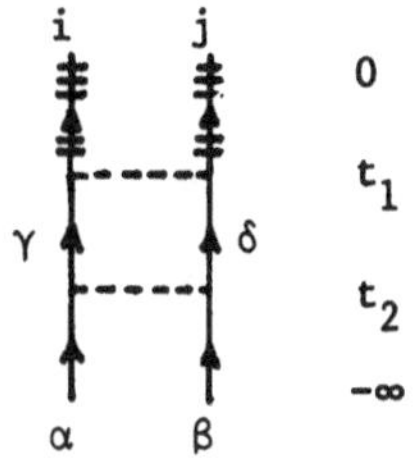

A

which consists of an infinite collection of wave function diagrams, among which is the diagram A on the far r.h.s. of eq. (6). Here, we use the time-ordered diagram, corresponding to using eq. (5) for $U(0,-\infty)$. Diagram A has the following meaning: It is an $n = 2$ term; i.e. there are two interactions (for convenience we take $H_1 = V$ instead of $H_1 = V - U$) acting successively at the times t_2 and t_1. At t_2 the two valence particles are scatttered into the valence states γ and δ, and at t_1 further scattered into the passive particle states i and j. The value of diagram A is

$$A = a_i^\dagger a_j^\dagger |c> \times \lim_{t' \to -\infty(\varepsilon)} \left(\frac{-i}{\hbar}\right)^2 \int_{t'}^{0} dt_1 \int_{t'}^{t_1} dt_2 \, e^{-i(\varepsilon_\gamma + \varepsilon_\delta - \varepsilon_i - \varepsilon_j)t_1/\hbar}$$

$$\times \, e^{-i(\varepsilon_\alpha + \varepsilon_\beta - \varepsilon_\gamma - \varepsilon_\delta)t_2/\hbar} \, \frac{1}{2} V_{ij,\gamma\delta} \frac{1}{2} V_{\gamma\delta,\alpha\beta} \ . \tag{6.1}$$

Recall that in the interaction representation the propagators at a given vertex (at time t) are collected according to the rule

$$e^{-i(\Sigma \varepsilon_{in} - \Sigma \varepsilon_{out})t/\hbar} \ . \tag{6.2}$$

Further, we have expressed the interaction V in terms of non-antisymmetrized matrix elements

$$V = \frac{1}{2} \sum_{k\ell mn} V_{k\ell,mn} \, a_k^\dagger a_\ell^\dagger a_n a_m \ , \tag{6.3}$$

and hence we include a factor 1/2 at each vertex. In the end we shall change to topologically distinct Hugenholtz diagrams, corresponding to using antisymmetrized vertices and no factors. Note also that in eq. (6.1) there is no time factor associated with the initial state $a_\alpha^\dagger a_\beta^\dagger|0>$, in which the system is at the time $-\infty$. In the Schrödinger representation, the time development of state vectors is given by $\Psi(t) = \exp[-iH(t-t')/\hbar]\Psi(t')$, while in the interaction representation the time dependence is $\Psi_I(t) = \exp[iH_o t/\hbar]\Psi(t)$, where $\Psi(t)$ is a Schrödinger state vector. Thus, if the unperturbed system is in a Schrödinger state $\Phi_a = a_\alpha^\dagger a_\beta^\dagger|0>$ at the time t_1, then at the time t_2 it is in the interaction state

$$\Phi_{a,I}(t_2) = e^{iH_o t_2/\hbar} \Phi_a(t_2) = e^{iH_o t_2/\hbar} e^{-iH_o(t_2-t_1)/\hbar} \Phi_a \ , \tag{7}$$

which is independent of t_2, i.e. time independent. It is customary as well as convenient to choose $t_1 = 0$ in eq. (7). We have employed this choice throughout, and hence we do not have any time factor associated with $a_\alpha^\dagger a_\beta^\dagger|0>$.

We can now rewrite diagram A as

$$A = a_i^\dagger a_j^\dagger |c\rangle \times \frac{1}{4} V_{ij,\gamma\delta} V_{\gamma\delta,\alpha\beta} I(A) , \tag{8}$$

where I(A) is the time integral

$$I(A) = \lim_{t' \to -\infty(\varepsilon)} \left(\frac{-i}{\hbar}\right)^2 \int_{t'}^{0} dt_1 \int_{t'}^{t_1} dt_2 \, e^{-i(\varepsilon_\gamma+\varepsilon_\delta-\varepsilon_i-\varepsilon_j)t_1/\hbar} \, e^{-i(\varepsilon_\alpha+\varepsilon_\beta-\varepsilon_\gamma-\varepsilon_\delta)t_2/\hbar} . \tag{8.1}$$

We shall frequently use a degenerate or nearly degenerate model space. In this case we have $\varepsilon_\alpha+\varepsilon_\beta-\varepsilon_\gamma-\varepsilon_\delta \approx 0$, and the integral I(A) is divergent. In order to handle this difficulty, we shall factorize out such terms in $U(0,-\infty)|\Phi_i\rangle$. With this in mind, we can rewrite diagram A as follows

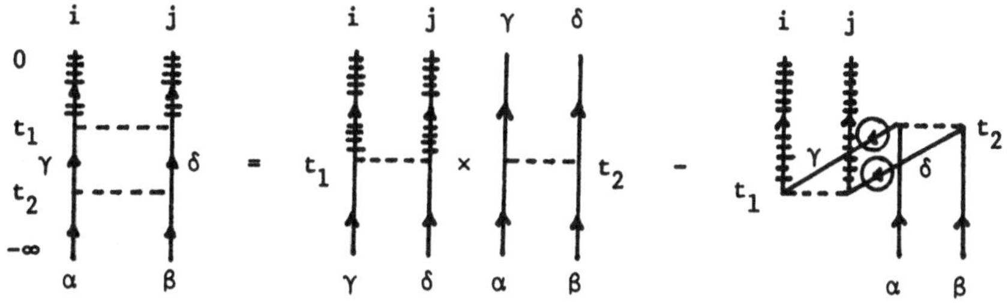

(9)

$0>t_1>t_2>-\infty$ $0>t_1>-\infty$, $0>t_2>-\infty$ $0>t_2>t_1>-\infty$

A   B   C

Here, diagram B represents a factorization of diagram A into two independent pieces. Thus, we have for diagram B

$$B = a_i^\dagger a_j^\dagger |c\rangle \times \frac{1}{4} V_{ij,\gamma\delta} V_{\gamma\delta,\alpha\beta} I(B) , \tag{9.1}$$

where the time integral I(B) is given by

$$I(B) = \lim_{t' \to -\infty(\varepsilon)} \left(\frac{-i}{\hbar}\right)^2 \int_{t'}^{0} dt_1 \, e^{-i(\varepsilon_\gamma+\varepsilon_\delta-\varepsilon_i-\varepsilon_j)t_1/\hbar} \int_{t'}^{0} dt_2 \, e^{-i(\varepsilon_\alpha+\varepsilon_\beta-\varepsilon_\gamma-\varepsilon_\delta)t_2/\hbar} . \tag{9.2}$$

Here, factorization implies that the two time integrations are independent of each other. Clearly diagram B is divergent, just like diagram A. However, diagrams A and B are not equal, because the time limits are different. Thus, diagram B contains a time-incorrect contribution, which is represented by the folded diagram C in eq. (9). Clearly, diagram C is given by

$$C = B - A = a_i^\dagger a_j^\dagger |c\rangle \times \frac{1}{4} V_{ij,\gamma\delta} V_{\gamma\delta,\alpha\beta} I(C) , \tag{9.3}$$

where the time integral $I(C)$ is

$$I(C) = I(B) - I(A) = \lim_{t' \to -\infty(\varepsilon)} \left(\frac{-i}{\hbar}\right)^2 \int_{t'}^{0} dt_1 \int_{t_1}^{0} dt_2 \, e^{-i(\varepsilon_\gamma+\varepsilon_\delta-\varepsilon_i-\varepsilon_j)t_1/\hbar} e^{-i(\varepsilon_\alpha+\varepsilon_\beta-\varepsilon_\gamma-\varepsilon_\delta)t_2/\hbar} . \tag{9.4}$$

It is interesting to note that $I(C)$ is finite even though $I(A)$ and $I(B)$ are infinite. In order to evaluate $I(C)$, we rewrite the integration limits as follows

$$\int_{t'}^{0} dt_1 \int_{t_1}^{0} dt_2 = \int_{t'}^{0} dt_2 \int_{t'}^{t_2} dt_1 . \tag{9.5}$$

Clearly, the areas of integration on the l.h.s. and r.h.s. coincide, as shown in fig. 2. In fact, the time limits on the r.h.s. can be read directly off diagram C in eq. (9). Substituting eq. (9.5) into eq. (9.4), we have

$$I(C) = \lim_{t' \to -\infty(\varepsilon)} \left(\frac{-i}{\hbar}\right)^2 \int_{t'}^{0} dt_2 \int_{t'}^{t_2} dt_1 \, e^{-i(\varepsilon_\gamma+\varepsilon_\delta-\varepsilon_i-\varepsilon_j)t_1/\hbar} e^{-i(\varepsilon_\alpha+\varepsilon_\beta-\varepsilon_\gamma-\varepsilon_\delta)t_2/\hbar}$$

$$= \lim_{t' \to -\infty(\varepsilon)} \left(\frac{-i}{\hbar}\right) \int_{t'}^{0} dt_2 \, e^{-i(\varepsilon_\alpha+\varepsilon_\beta-\varepsilon_\gamma-\varepsilon_\delta)t_2/\hbar} \frac{1}{\varepsilon_\gamma+\varepsilon_\delta-\varepsilon_i-\varepsilon_j}$$

$$\times \{e^{-i(\varepsilon_\gamma+\varepsilon_\delta-\varepsilon_i-\varepsilon_j)t_2/\hbar} - e^{-i(\varepsilon_\gamma+\varepsilon_\delta-\varepsilon_i-\varepsilon_j)t'/\hbar}\}$$

$$= \frac{1}{\varepsilon_\gamma+\varepsilon_\delta-\varepsilon_i-\varepsilon_j} \lim_{t'\to -\infty(\varepsilon)} \left(\frac{-i}{\hbar}\right) \int_{t'}^{0} dt_2 \, e^{-i(\varepsilon_\alpha+\varepsilon_\beta-\varepsilon_i-\varepsilon_j)t_2/\hbar}$$

$$= \frac{1}{\varepsilon_\gamma+\varepsilon_\delta-\varepsilon_i-\varepsilon_j} \, \frac{1}{\varepsilon_\alpha+\varepsilon_\beta-\varepsilon_i-\varepsilon_j} \; . \tag{9.6}$$

In the above integral all terms of the form $\exp[-i(\varepsilon_\alpha+\varepsilon_\beta-\varepsilon_i-\varepsilon_j)t'/\hbar]$ vanish in the limit $t'\to -\infty(\varepsilon)$, since $\varepsilon_\alpha+\varepsilon_\beta-\varepsilon_i-\varepsilon_j < 0$. Thus, I(C) becomes finite.

Altogether, we obtain for diagram C (not including the minus sign in front)

$$C = a_i^\dagger a_j^\dagger \, |c\rangle \times \frac{1}{4} V_{ij,\gamma\delta} V_{\gamma\delta,\alpha\beta} \frac{1}{(\varepsilon_\gamma+\varepsilon_\delta-\varepsilon_i-\varepsilon_j)(\varepsilon_\alpha+\varepsilon_\beta-\varepsilon_i-\varepsilon_j)} \; . \tag{9.7}$$

This expression can be further simplified if the model space is assumed to be degenerate. In this case we obtain a squared energy denominator $(\varepsilon_\gamma+\varepsilon_\delta-\varepsilon_i-\varepsilon_j)^{-2}$. In general, the calculation of folded diagrams is often simplified by assuming a degenerate model space, and hence we shall frequently make this assumption from now on.

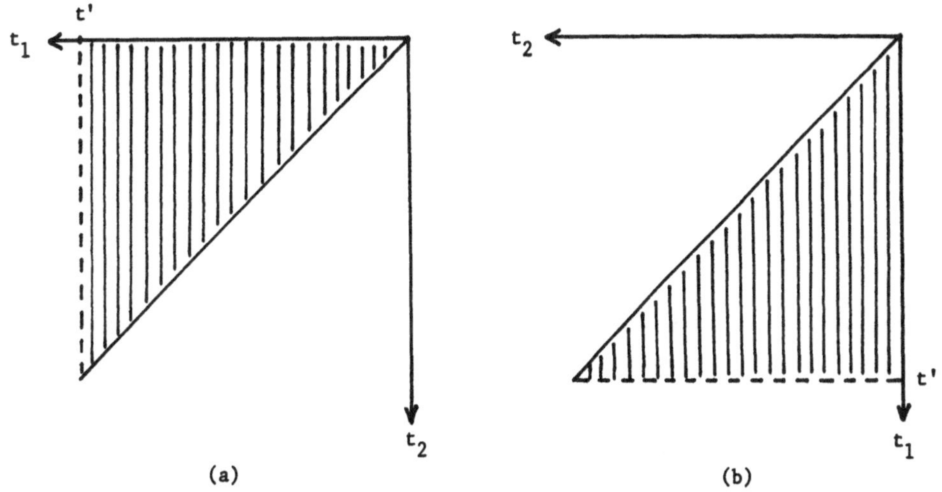

Fig. 2. Integration areas on (a) the l.h.s. and (b) the r.h.s. of eq. (9.5).

The rules for evaluating folded diagrams are simple. In diagram C the folded lines γ and δ refer to particle states, but are drawn as downward going lines (with a circle to distinguish them from real hole lines). In fact, we can use the ordinary diagram rules if the folded lines are treated as hole lines in evaluating the energy denominator

$$\frac{1}{(\varepsilon_\alpha+\varepsilon_\beta)-(\varepsilon_\alpha+\varepsilon_\beta+\varepsilon_i+\varepsilon_j-\varepsilon_\gamma-\varepsilon_\delta)}\,\frac{1}{(\varepsilon_\alpha+\varepsilon_\beta)-(\varepsilon_i+\varepsilon_j)} = \frac{1}{(\varepsilon_\gamma+\varepsilon_\delta-\varepsilon_i-\varepsilon_j)(\varepsilon_\alpha+\varepsilon_\beta-\varepsilon_i-\varepsilon_j)}\,, \tag{9.8}$$

which agrees with eq. (9.6). This is the only modification of the diagram rules needed to evaluate folded diagrams (not including the sign in front; this sign may be obtained by associating a factor (-1) with each set of folded lines).

There are more complicated folded diagrams, as shown by the following example

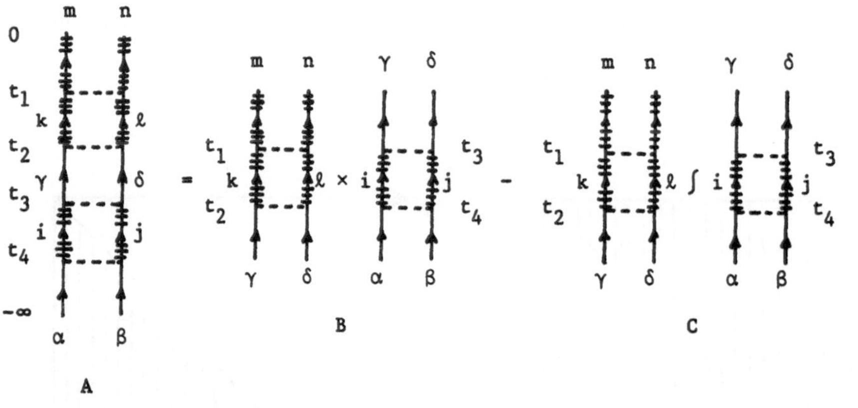

(10)

All the diagrams A, B and C have identical integrands and constant factors, namely

$$a_m^\dagger a_n^\dagger |c\rangle \times \left(\tfrac{1}{2}\right)^4 V_{mn,k\ell}\, V_{k\ell,\gamma\delta}\, V_{\gamma\delta,ij}\, V_{ij,\alpha\beta} \left(\tfrac{-i}{\hbar}\right)^4 e^{-i(\varepsilon_k+\varepsilon_\ell-\varepsilon_m-\varepsilon_n)t_1/\hbar}$$

$$\times\, e^{-i(\varepsilon_\gamma+\varepsilon_\delta-\varepsilon_k-\varepsilon_\ell)t_2/\hbar}\, e^{-i(\varepsilon_i+\varepsilon_j-\varepsilon_\gamma-\varepsilon_\delta)t_3/\hbar}\, e^{-i(\varepsilon_\alpha+\varepsilon_\beta-\varepsilon_i-\varepsilon_j)t_4/\hbar}\,. \tag{10.1}$$

The only difference is in the integration limits. The integration limits of diagram A are given by

$$\int_{t'}^{0} dt_1 \int_{t'}^{t_1} dt_2 \int_{t'}^{t_2} dt_3 \int_{t'}^{t_3} dt_4 \, , \tag{10.2}$$

corresponding to the time ordering $0 > t_1 > t_2 > t_3 > t_4 > t'$. This time ordering is violated when we factorize diagram A into two independent pieces, as shown in diagram B. The time ordering of diagram B is $0 > t_1 > t_2 > t'$ and $0 > t_3 > t_4 > t'$, and the corresponding limits of integration are

$$\int_{t'}^{0} dt_1 \int_{t'}^{t_1} dt_2 \int_{t'}^{0} dt_3 \int_{t'}^{t_3} dt_4 \, . \tag{10.3}$$

To correct for the incorrect time ordering in diagram B, we have introduced the folded diagram C, which has the time constraints $0 > t_1 > t_2 > t'$, $0 > t_3 > t_4 > t'$ and $t_3 > t_2$. There are in fact five different time sequences satisfying these constraints. Thus, diagram C, which will be called a generalized folded diagram (the "integral" sign denoting generalized folding) actually consists of five ordinary folded diagrams

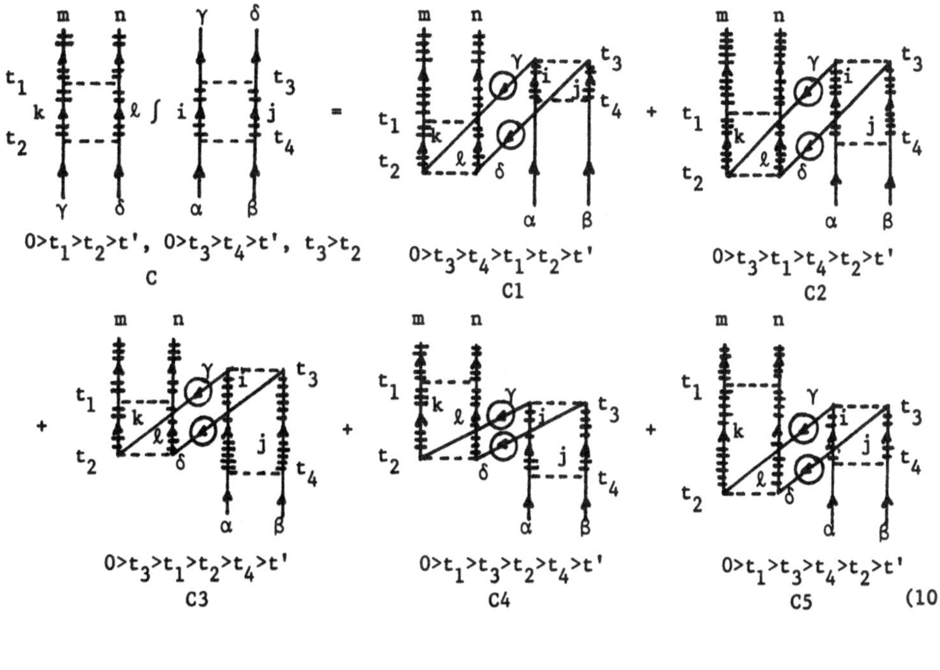

(10.4)

Each of the diagrams C1-C5 has a specific time ordering. Hence, their time integrals are easy to evaluate. For example, the time integral I(C2) of diagram C2 is

$$I(C2) = \lim_{t' \to -\infty(\varepsilon)} \left(\frac{-i}{\hbar}\right)^4 \int_{t'}^{0} dt_3 \int_{t'}^{t_3} dt_1 \int_{t'}^{t_1} dt_4 \int_{t'}^{t_4} dt_2 \; e^{-i(\varepsilon_i+\varepsilon_j-\varepsilon_\gamma-\varepsilon_\delta)t_3/\hbar}$$

$$\times e^{-i(\varepsilon_k+\varepsilon_\ell-\varepsilon_m-\varepsilon_n)t_1/\hbar} \; e^{-i(\varepsilon_\alpha+\varepsilon_\beta-\varepsilon_i-\varepsilon_j)t_4/\hbar} \; e^{-i(\varepsilon_\gamma+\varepsilon_\delta-\varepsilon_k-\varepsilon_\ell)t_2/\hbar}$$

$$= \frac{1}{\varepsilon_\gamma+\varepsilon_\delta-\varepsilon_k-\varepsilon_\ell} \lim_{t' \to -\infty(\varepsilon)} \left(\frac{-i}{\hbar}\right)^3 \int_{t'}^{0} dt_3 \int_{t'}^{t_3} dt_1 \int_{t'}^{t_1} dt_4 \; e^{-i(\varepsilon_i+\varepsilon_j-\varepsilon_\gamma-\varepsilon_\delta)t_3/\hbar}$$

$$\times e^{-i(\varepsilon_k+\varepsilon_\ell-\varepsilon_m-\varepsilon_n)t_1/\hbar} \; e^{-i(\varepsilon_\alpha+\varepsilon_\beta+\varepsilon_\gamma+\varepsilon_\delta-\varepsilon_i-\varepsilon_j-\varepsilon_k-\varepsilon_\ell)t_4/\hbar}$$

$$= \; \ldots\ldots$$

$$= \frac{1}{\varepsilon_\gamma+\varepsilon_\delta-\varepsilon_k-\varepsilon_\ell} \frac{1}{(\varepsilon_\alpha+\varepsilon_\beta+\varepsilon_\gamma+\varepsilon_\delta)-(\varepsilon_i+\varepsilon_j+\varepsilon_k+\varepsilon_\ell)}$$

$$\times \frac{1}{(\varepsilon_\alpha+\varepsilon_\beta+\varepsilon_\gamma+\varepsilon_\delta)-(\varepsilon_i+\varepsilon_j+\varepsilon_m+\varepsilon_n)} \frac{1}{\varepsilon_\alpha+\varepsilon_\beta-\varepsilon_m-\varepsilon_n} \; . \qquad (10.5)$$

At each integration the contribution from t' vanishes in the complex-time limit since the exponent is negative.

Obviously, the energy denominator between two successive vertices in a time-ordered folded diagram is given by the following rule

$$\{ \sum_s \varepsilon_s - \sum_p \varepsilon_p + \sum_h \varepsilon_h + \sum_f \varepsilon_f \}^{-1} , \qquad (11)$$

where s represents the incoming lines, and p, h and f denote the particle, hole and folded lines in the interval concerned.

The generalized folded diagram C can be calculated in two ways. Clearly, one can calculate all the individual time-ordered folded diagrams and add them up. However, this will easily become very cumbersome when dealing with higher order folded diagrams, since there are so many of them. Alternatively, one can calculate diagram C as the difference between diagrams B and A. The latter approach is more convenient for higher order diagrams, and will be used here. We shall only be concerned with the time integrals (i.e. energy denominators) since the other factors are identical for the three diagrams. By straightforward application of the energy-denominator rule $\{\sum_s \varepsilon_s - \sum_p \varepsilon_p + \sum_h \varepsilon_h\}^{-1}$ for non-folded valence-particle diagrams we obtain for diagrams A and B

$$I(A) = \frac{1}{(\varepsilon_\alpha+\varepsilon_\beta-\varepsilon_m-\varepsilon_n)}\,\frac{1}{(\varepsilon_\alpha+\varepsilon_\beta-\varepsilon_k-\varepsilon_\ell)}\,\frac{1}{(\varepsilon_\alpha+\varepsilon_\beta-\varepsilon_\gamma-\varepsilon_\delta)}\,\frac{1}{(\varepsilon_\alpha+\varepsilon_\beta-\varepsilon_i-\varepsilon_j)} \tag{11.1}$$

and

$$I(B) = \frac{1}{(\varepsilon_\gamma+\varepsilon_\delta-\varepsilon_m-\varepsilon_n)}\,\frac{1}{(\varepsilon_\gamma+\varepsilon_\delta-\varepsilon_k-\varepsilon_\ell)} \times \frac{1}{(\varepsilon_\alpha+\varepsilon_\beta-\varepsilon_\gamma-\varepsilon_\delta)}\,\frac{1}{(\varepsilon_\alpha+\varepsilon_\beta-\varepsilon_i-\varepsilon_j)} \,. \tag{11.2}$$

Taking the difference, we get

$$I(C) = I(B) - I(A) = \frac{1}{\varepsilon_\alpha+\varepsilon_\beta-\varepsilon_\gamma-\varepsilon_\delta}\left\{\frac{1}{(\varepsilon_\gamma+\varepsilon_\delta-\varepsilon_m-\varepsilon_n)(\varepsilon_\gamma+\varepsilon_\delta-\varepsilon_k-\varepsilon_\ell)} - \frac{1}{(\varepsilon_\alpha+\varepsilon_\beta-\varepsilon_m-\varepsilon_n)(\varepsilon_\alpha+\varepsilon_\beta-\varepsilon_k-\varepsilon_\ell)}\right\}\frac{1}{\varepsilon_\alpha+\varepsilon_\beta-\varepsilon_i-\varepsilon_j} \,. \tag{11.3}$$

In the degenerate case, in which we shall often be interested, the first factor is infinite, while the second factor is zero. Thus, I(C) has to be determined by a limiting procedure. Writing $\varepsilon_\alpha + \varepsilon_\beta = \varepsilon_\gamma + \varepsilon_\delta + \Delta$, where $\Delta \to 0$, we can cast eq. (11.3) into the form

$$I(C) = \lim_{\Delta\to 0} \frac{1}{\Delta}\left\{\frac{1}{XY} - \frac{1}{(X+\Delta)(Y+\Delta)}\right\}\frac{1}{Z+\Delta} = \left\{\frac{1}{X^2Y} + \frac{1}{XY^2}\right\}\frac{1}{Z} \,, \tag{11.4}$$

where

$$X = \varepsilon_\gamma + \varepsilon_\delta - \varepsilon_m - \varepsilon_n \,,$$
$$Y = \varepsilon_\gamma + \varepsilon_\delta - \varepsilon_k - \varepsilon_\ell \,,$$
$$Z = \varepsilon_\gamma + \varepsilon_\delta - \varepsilon_i - \varepsilon_j \,. \tag{11.5}$$

Thus, we can express the energy denominator of the generalized folded diagram C as the derivative of the energy denominator of the left-hand part of the diagram with respect to the starting energy

$$I(C) = -\frac{d}{d\omega}\left\{\frac{1}{\omega-\varepsilon_k-\varepsilon_\ell}\;\frac{1}{\omega-\varepsilon_m-\varepsilon_n}\right\}_{\omega=\varepsilon_\gamma+\varepsilon_\delta} \times \frac{1}{\varepsilon_\gamma+\varepsilon_\delta-\varepsilon_i-\varepsilon_j} \,. \tag{11.6}$$

We shall come back to this derivative method later, as it is a convenient scheme for computing folded diagrams in a degenerate model space.

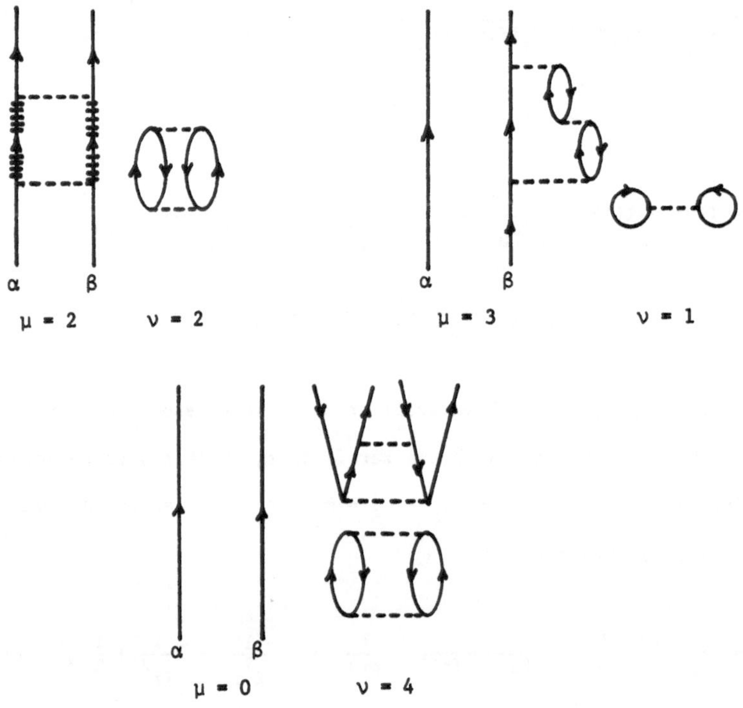

Fig. 3. Fourth order diagrams classified according to the number of vertices attached to valence lines.

3. The decomposition theorem

The aim of the folded-diagram factorization procedure developed in the previous section is to factorize $U(0,-\infty)|\Phi_i\rangle$ as a whole. In eq. (6), which we repeat here for convenience

$$U(0,-\infty)|\Phi_i\rangle = U(0,-\infty)\, a_\alpha^\dagger\, a_\beta^\dagger\, |c\rangle , \tag{12}$$

where $U(0,-\infty)$ is given by eq. (5.1), we divide the n H_1-vertices of a given n-th order term into two groups:

(i) one group consisting of μ vertices linked (directly or indirectly) to one or both of the valence lines α and β, and

(ii) one group consisting of $\nu = n - \mu$ vertices which are not linked to any of the valence lines.

Examples of such (μ,ν) partitions of various fourth order terms are shown in fig. 3. There are clearly $\frac{n!}{\mu!\nu!}$ different ways to make a (μ,ν) partition of n objects. Since all the different possibilities of making a given (μ,ν) partition give the same contribution to $U(0,-\infty)$ and furthermore, since

$$\sum_{n=0}^{\infty} \frac{1}{n!} \sum_{\mu+\nu=n}^{\infty} \frac{n!}{\mu!\nu!} = \sum_{\mu=0}^{\infty} \frac{1}{\mu!} \sum_{\nu=0}^{\infty} \frac{1}{\nu!} , \tag{12.1}$$

we obtain from eq. (5.1)

$$U(0,-\infty)\, a_\alpha^\dagger\, a_\beta^\dagger\, |c\rangle = U_V(0,-\infty)\, a_\alpha^\dagger\, a_\beta^\dagger\, |c\rangle \times U(0,-\infty)|c\rangle . \tag{12.2}$$

Here, the subscript V (valence) indicates that all the interactions in $U_V(0,-\infty)$ are linked to one or both of the valence lines α and β. Typical diagrams contained in $U_V(0,-\infty)a_\alpha^\dagger a_\beta^\dagger|c\rangle$ are shown in fig. 4.

It is important to note that the factorization of eq. (12.2) is possible only if the Pauli exclusion principle is not enforced locally. We explain

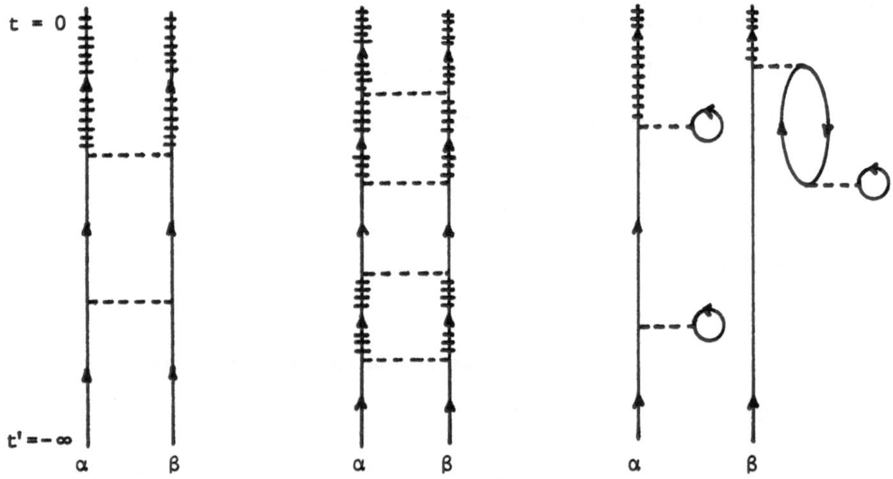

Fig. 4. Diagrams contained in $U_V(0,-\infty)a_\alpha^\dagger a_\beta^\dagger|c\rangle$. Note that all the vertices are linked (directly or indirectly) to at least one valence line.

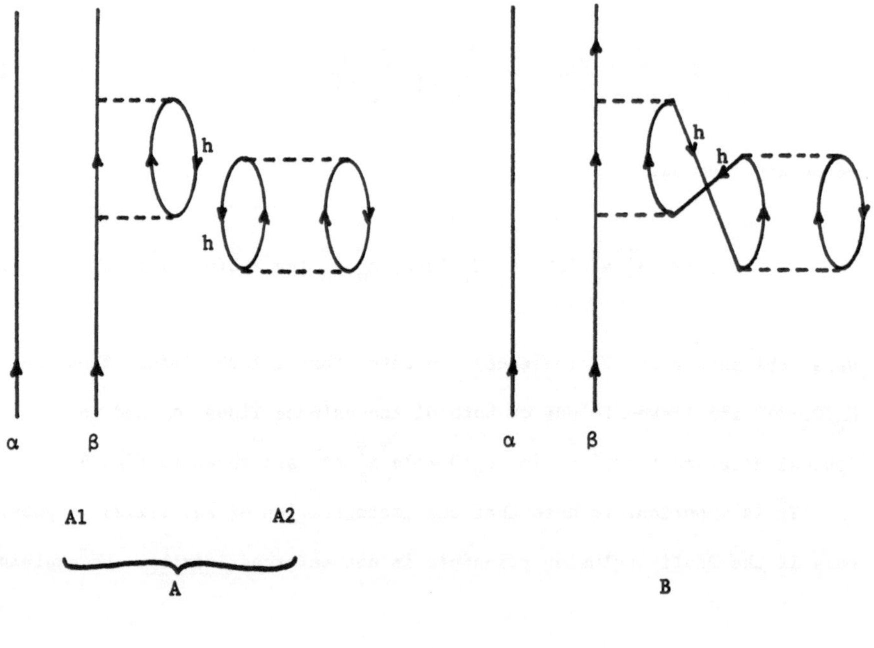

Fig. 5. Pauli-violating diagrams which cancel each other.

this by way of an example. Consider the two diagrams shown in fig. 5, both of which belong to $U(0,-\infty)a_\alpha^\dagger a_\beta^\dagger|c>$. The Pauli exclusion principle is locally violated in both A and B. But since $A + B = 0$, the Pauli exclusion principle is globally observed. Diagram A is the product of two diagrams A1 and A2. In the factorization scheme of eq. (12.2), A1 is included in $U_V a_\alpha^\dagger a_\beta^\dagger|c>$ and A2 in $U|c>$. Thus, to preserve the Pauli exclusion principle globally, we must include diagram B in $U_V a_\alpha^\dagger a_\beta^\dagger|c>$.

To summarize, in $U_V a_\alpha^\dagger a_\beta^\dagger|c>$ we include the diagrams in which all the vertices are linked to at least one valence line, including those which violate the Pauli exclusion principle.

Then, we proceed to factorize each of the terms on the r.h.s. of eq. (12.2). Consider first the second term. Using the expansion of eq. (5.1), we write

$$U(0,-\infty)|c> = \sum_{n=0}^{\infty} U^{(n)}(0,-\infty)|c>, \tag{12.3}$$

where $U^{(n)}(0,-\infty)|c>$ represents all the terms with n vertices of H_1. The $n = 0$ term is just $|c>$. We can divide the n vertices of $U^{(n)}(0,-\infty)|c>$ into two groups: One has μ vertices which are all connected, directly or indirectly, to the $t = 0$ time boundary, and the other has $\nu = n - \mu$ vertices which do not connect to this boundary. Some examples of this division are shown in fig. 6. Now we group the various diagrams using a procedure almost identical to that leading to eq. (12.2). There are $\frac{n!}{\mu!\nu!}$ equivalent ways of making the aforementioned (μ,ν) partition. Then, using eq. (12.1), we can rewrite eq. (12.3) as

$$U(0,-\infty)|c> = U_Q(0,-\infty)|c><c|U(0,-\infty)|c>, \tag{13}$$

Here, $U_Q|c>$ originates from the first group of vertices (μ) and thus, in addition to $|c>$, contains all the wave function diagrams in which

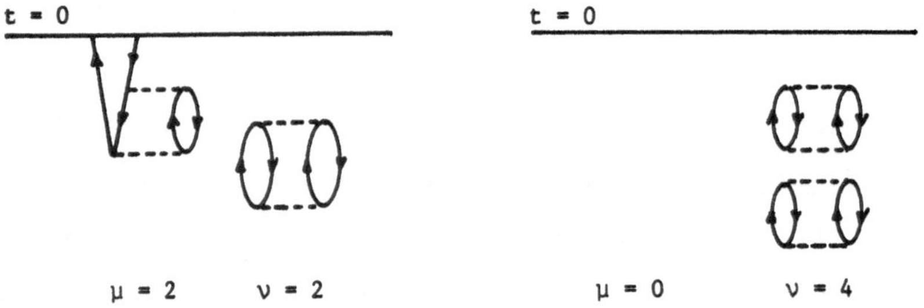

Fig. 6. Classification of diagrams contained in $U^{(4)}(0,-\infty)|c\rangle$.

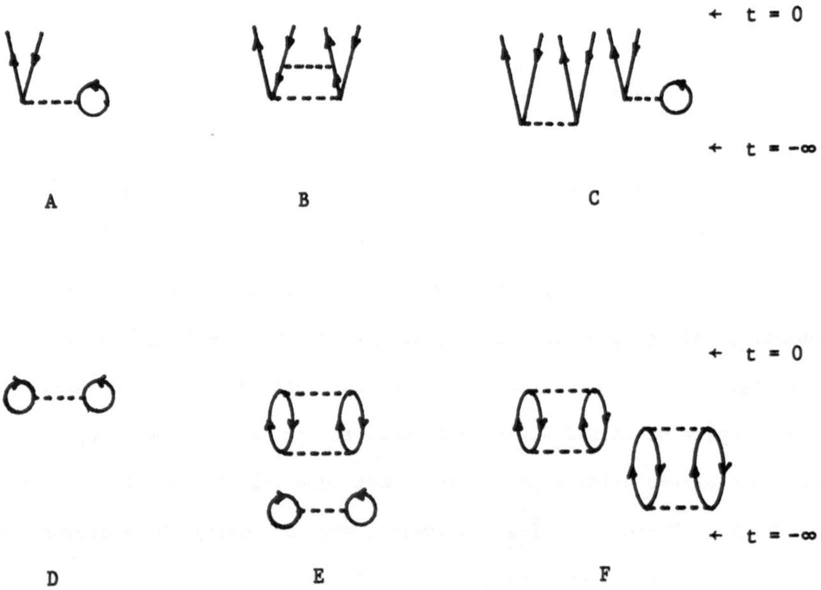

Fig. 7. Diagrams contained in $U_Q(0,-\infty)|c\rangle$ (A-C) and in $\langle c|U(0,-\infty)|c\rangle$ (D-F). Note that for the former all the vertices are connected to the time boundary $t = 0$, whereas for the latter they are not.

every vertex is connected to the time $t = 0$ via fermion lines. In fact, $U_Q|c\rangle$ is proportional to the true ground-state wave function $|\Psi_o^c\rangle$ of the core system[25]. On the other hand, $\langle c|U|c\rangle$ originates from the second group of vertices (ν) and represents all the vacuum fluctuation diagrams. These classifications are illustrated in fig. 7.

We can perform a similar factorization of the first term on the r.h.s. of eq. (12.2). The diagrams contained in $U_V a_\alpha^\dagger a_\beta^\dagger|c\rangle$ can be expressed in terms of so-called $\hat{Q}$-boxes. The $\hat{Q}$-box is defined as[25)]

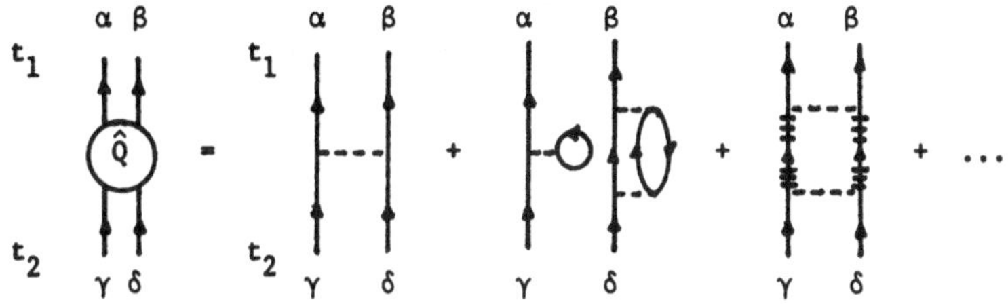

(14)

which is the sum of all the diagrams (between the times t_1 and $t_2 < t_1$) that have <u>at least one H_1-vertex</u> and are <u>irreducible</u> (i.e. contain at least one passive line between two successive vertices) and <u>valence-linked</u> (i.e. have all their vertices linked to at least one active (valence) line). For example, the diagram shown in fig. 8 is not allowed in the $\hat{Q}$-box because it has two vertices which are not linked to any valence line. We shall discuss the rules for evaluating the diagrams in the $\hat{Q}$-box in sect. 6. Here, we are interested in the factorization of the chains of $\hat{Q}$-boxes shown in eqs. (16.1-2) below.

In eq. (14) the $\hat{Q}$-box was defined between incoming and outgoing valence-particle lines. As we shall shortly see, we shall also need to define a $\hat{Q}$-box between incoming valence-particle (active) lines and outgoing passive lines. This $\hat{Q}$-box is defined in complete analogy with eq. (14) as

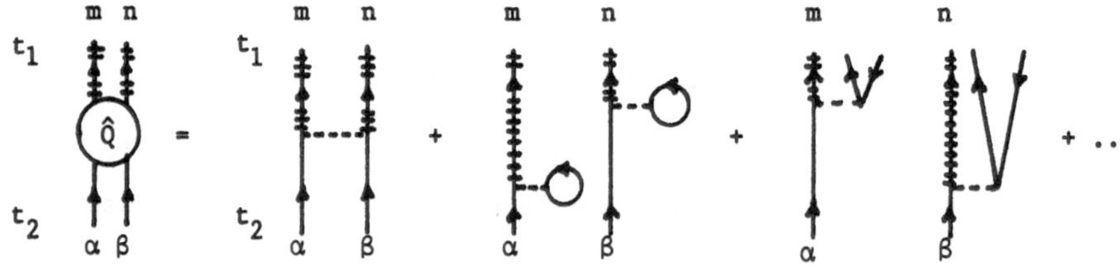

(14.1)

In order to further simplify (and generalize) the diagrams to be used in the following, we define an active state as being solely composed of valence-particle (active) lines and a passive state as being composed of at least one passive line. We introduce the following general notation for active and passive states:

Active state 

Passive state 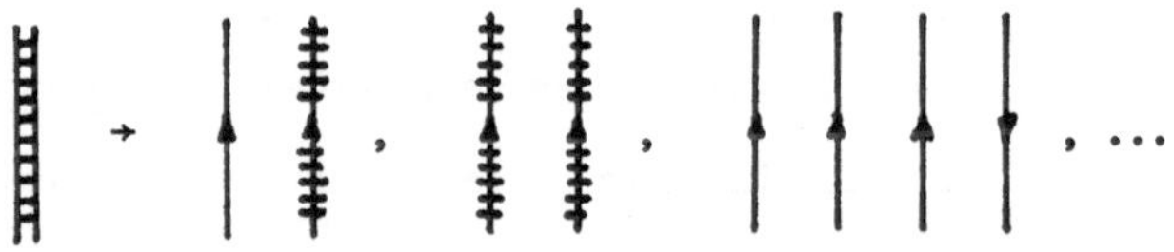 (15)

According to the definitions (14-14.1) we always have a passive state between two successive vertices in a $\hat{Q}$-box diagram.

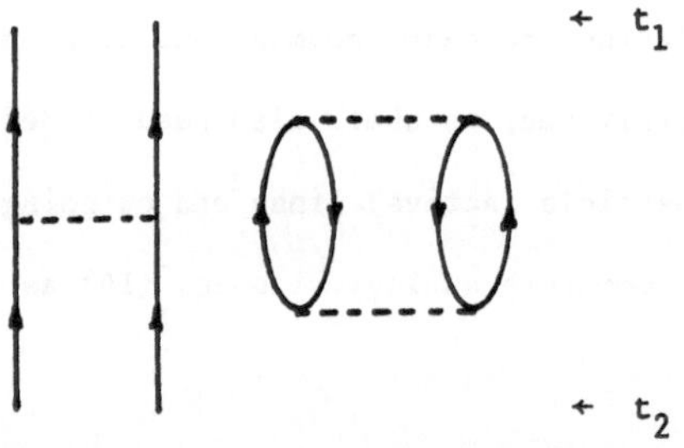

Fig. 8. Example of diagram which is not allowed in the $\hat{Q}$-box.

Then, we turn to the decomposition of $U_V(0,-\infty)a_\alpha^\dagger a_\beta^\dagger|c\rangle$ in terms of $\hat{Q}$-boxes. It is convenient to use the notation (15) which applies to the general case $U_V(0,-\infty)|\Phi_i\rangle$, where $|\Phi_i\rangle$ is a basis vector of an arbitrary d-dimensional P-space. Clearly, $U_V(0,-\infty)|\Phi_i\rangle$ must either terminate in an active or a passive state at $t = 0$. Hence, we can write

$$U_V(0,-\infty)|\Phi_i\rangle = |\chi_i^P\rangle + |\chi_i^Q\rangle , \tag{16}$$

where $|\chi_i^P\rangle$ terminates in an active state and $|\chi_i^Q\rangle$ in a passive state at $t = 0$. Thus, we have†

$|\chi_i^P\rangle = \sum_{j=1}^{d}$ 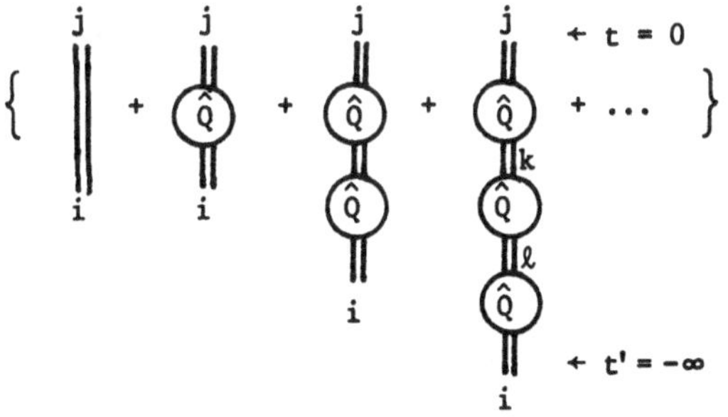

(16.1)

and

$|\chi_i^Q\rangle = \sum_{n>d}$ 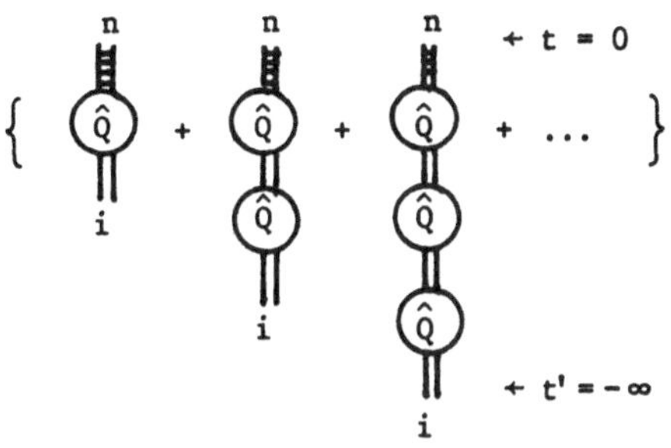

. (16.2)

---

† Note that we presently are considering wave function diagrams. Thus, we put $t = 0$ at the top of the diagram. For energy diagrams, we shall put $t = 0$ at the top of the highest $\hat{Q}$-box, since there will always be a last vertex of H_1 at $t = 0$, as we shall shortly see.

Here, summation over all P-space intermediate states is implied. For example, in the last term of eq. (16.1) both k and ℓ are summed from 1 to d . In eqs. (16.1-2) we have grouped the terms according to the number of active states between the interactions. Thus, there is no double counting.

We now factorize out of $|\chi_i^Q\rangle$ a factor belonging to $|\chi_i^P\rangle$ by means of folded-diagram factorization. Consider for example

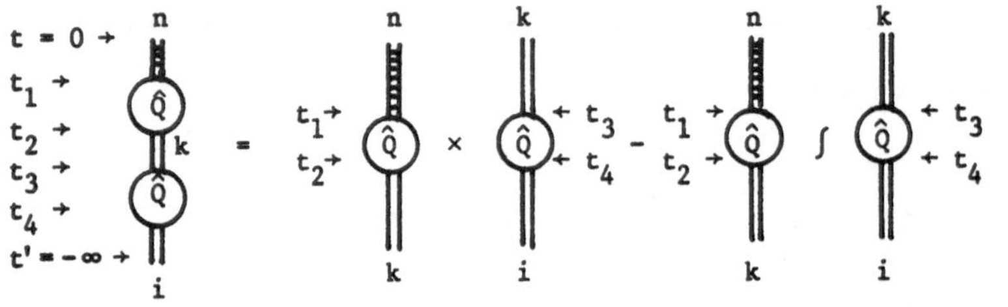

(16.3)

$0>t_1>t_2>t_3>t_4>t'$ $\quad$ $0>t_1>t_2>t'$, $0>t_3>t_4>t'$ $\quad$ $0>t_1>t_2>t'$, $0>t_3>t_4>t'$, $t_3>t_2$

A $\quad$ B $\quad$ C

Here, the upper and lower time limits on a $\hat{Q}$-box refer to the latest and earliest vertices, respectively, in the $\hat{Q}$-box. Note the similarity between eq. (16.3) and eqs. (9-10). Diagram A in eq. (16.3) has time constraints $t_1 > t_2$, $t_3 > t_4$ and $t_2 > t_3$. By factorizing A as shown in B, we introduce time-incorrect contributions with $t_3 > t_2$. Hence, to compensate for this, we need to introduce the generalized $\hat{Q}$-box folded diagram C.

The meaning of eq. (16.3) should be clear. Each of the diagrams A, B and C represents a collection of diagrams having identical general structure. For example, if we include in the $\hat{Q}$-boxes the following terms

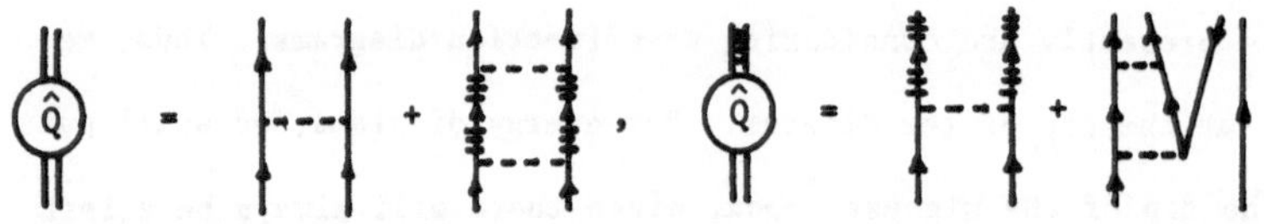

(16.4)

diagram C becomes a sum of four terms

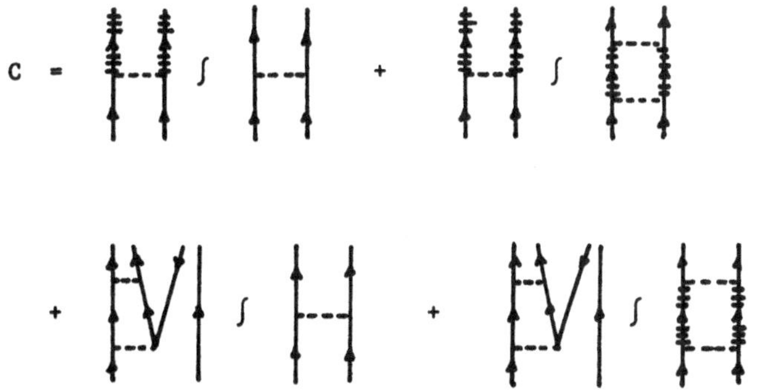

(16.5)

We can of course generalize the folding operation to diagrams with more than two $\hat{Q}$-boxes. For example, we have

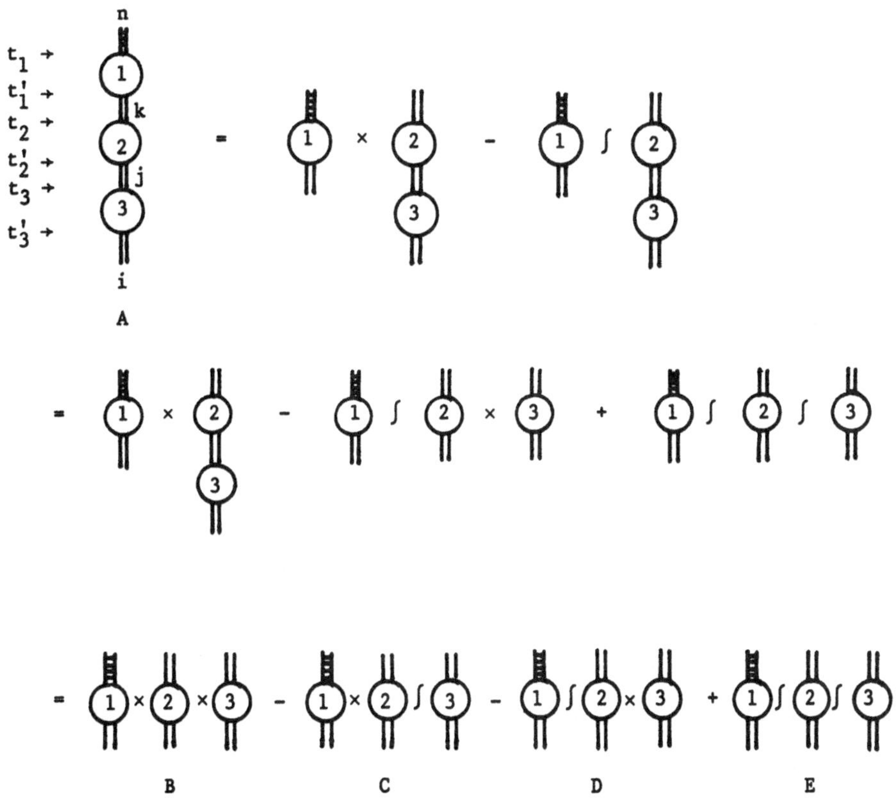

(16.6)

where for convenience we have numbered the $\hat{Q}$-boxes. Of interest here are the time limits for the various terms. Common to all the diagrams are the constraints

$$\text{A, B, C, D, E:} \qquad 0 > t_i > t_i' > -\infty, \qquad i = 1, 2, 3 . \tag{16.7}$$

These are the only constraints for the factorized diagram B. However, the original diagram A has the additional constraints

$$\text{A:} \qquad t_1' > t_2 \quad \text{and} \quad t_2' > t_3 . \tag{16.8}$$

Thus, to compensate for the incorrect time orderings in diagram B, we have to introduce the folded diagrams C, D and E with the additional constraints

$$\text{C:} \qquad t_3 > t_2' ,$$

$$\text{D:} \qquad t_2 > t_1' ,$$

$$\text{E:} \qquad t_2 > t_1' \quad \text{and} \quad t_3 > t_2' . \tag{16.9}$$

The general rule for the relative time constraints on two neighbouring $\hat{Q}$-boxes connected by a folding sign $\int$ is that the top time of the box to the right of $\int$ must be later than the bottom time of the box to the left.

Note that diagram E in eq. (16.6) is a twice-folded $\hat{Q}$-box diagram. Although we shall discuss the calculation of folded diagrams in detail later, it may be useful to give just one example now. We consider a time-ordered, twice-folded wave function diagram which is easily evaluated using the standard diagram rules (vertices not being antisymmetrized) supplemented with rule (11):

$0>t_1>t_2>t_3>t_4>-\infty$

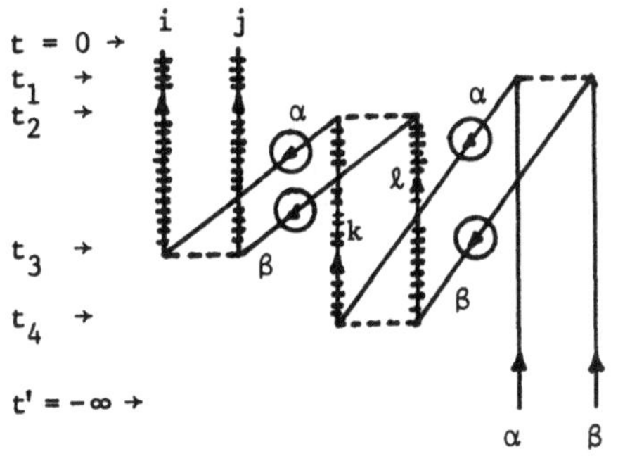

$$= a_i^\dagger a_j^\dagger |c\rangle (\tfrac{1}{2})^4 V_{ij,\alpha\beta} V_{\alpha\beta,k\ell} V_{k\ell,\alpha\beta} V_{\alpha\beta,\alpha\beta}$$

$$\times \frac{1}{(\varepsilon_\alpha+\varepsilon_\beta-\varepsilon_i-\varepsilon_j)^2} \frac{1}{2(\varepsilon_\alpha+\varepsilon_\beta)-(\varepsilon_i+\varepsilon_j+\varepsilon_k+\varepsilon_\ell)} \frac{1}{(\varepsilon_\alpha+\varepsilon_\beta-\varepsilon_k-\varepsilon_\ell)} . \qquad (17)$$

We are now ready to prove the decomposition theorem. Using the folding procedure of eqs. (16.3) and (16.6) we can readily factorize each term in $|\chi_i^Q\rangle$ to obtain a factor belonging to $|\chi_i^P\rangle$:

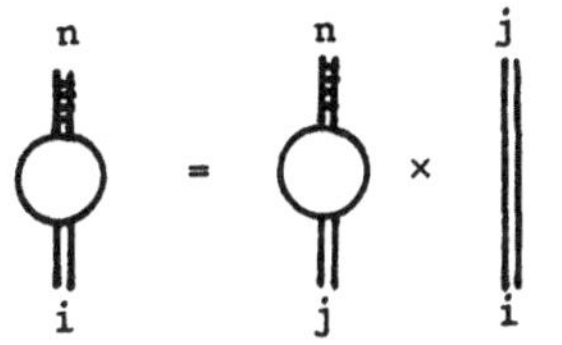

(18a)

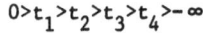

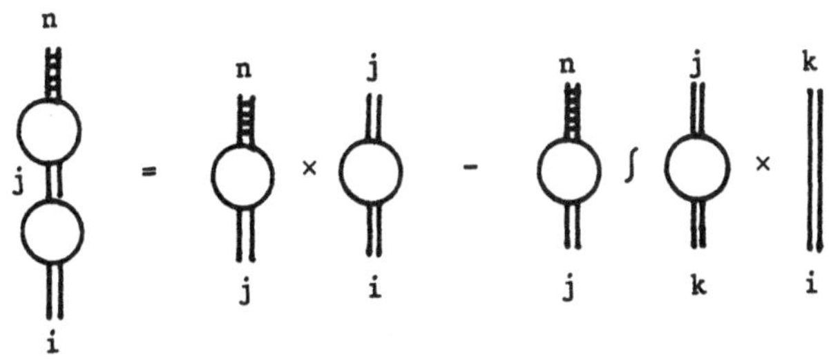

(18b)

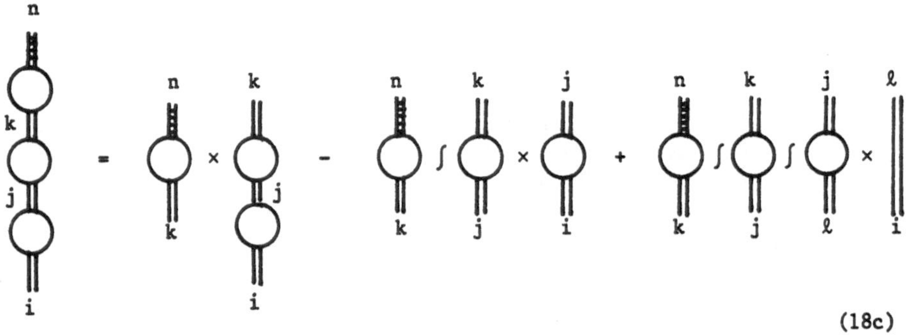

(18c)

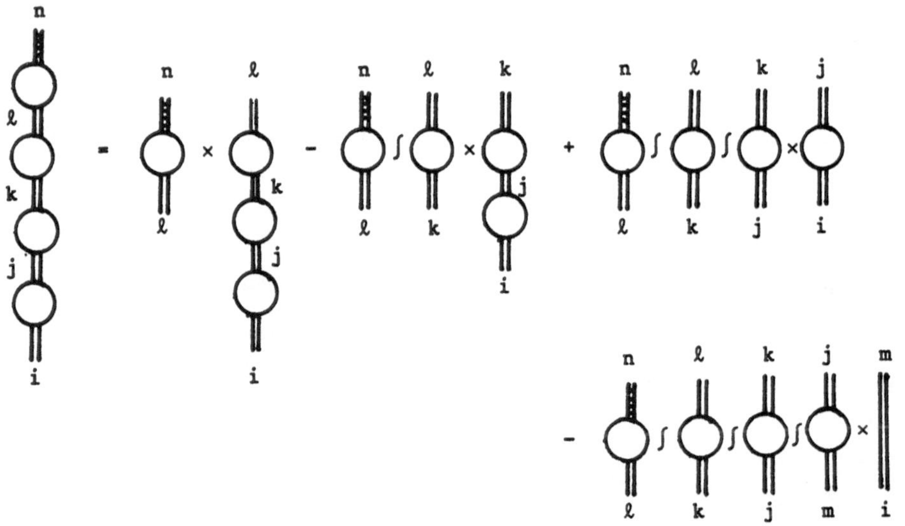

(18d)

etc.

Here, summations over all possible active intermediate states are implied. The second factor in eq. (18a) is just δ_{ij} and corresponds to the first term in $|\chi_i^P\rangle$. The second factor in the first term on the r.h.s. of eq. (18b) is the second term in $|\chi_i^P\rangle$. Similarly, we identify the corresponding factors in eqs. (18c) and (18d) with the third and fourth terms of $|\chi_i^P\rangle$, respectively, and so on and so forth. Then, we turn to the second term on the r.h.s. of eqs. (18b-d). The third factor in the second term of eq.

(18b) is the first term in $|\chi_i^P\rangle$, the third factor in the second term of eq. (18c) is the second term in $|\chi_i^P\rangle$, and so we can go on. Thus, by collecting the terms in the above equations columnwise and adding up, we obtain

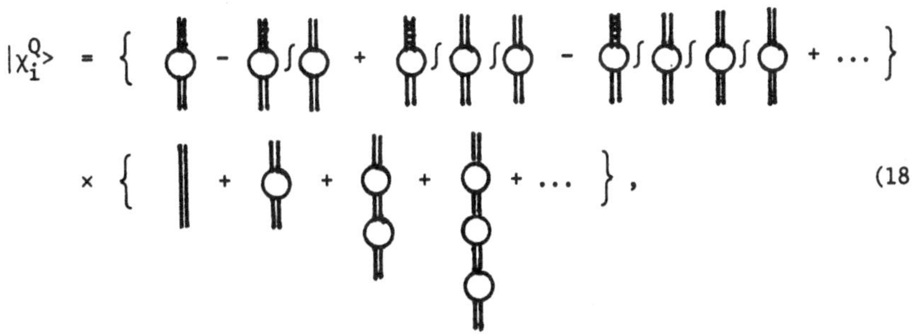

(18.1)

where the second bracket is related to $|\chi_i^P\rangle$ of eq. (16.1). To understand the precise meaning of the r.h.s. of eq. (18.1), we may treat the $\hat{Q}$-box as an operator $\hat{Q}(\omega)$ depending on the starting energy ω and write the diagrams in algebraic form. Then, we have for example

$$= |n\rangle\langle n|\hat{Q}(i)|i\rangle \tag{18.2}$$

and

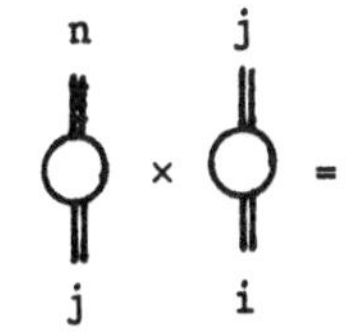

$$= \sum_j |n\rangle\langle n|\hat{Q}(j)|j\rangle\langle j|\hat{Q}(i)|i\rangle\,, \tag{18.3}$$

where we have used the notation $|i\rangle \equiv |\Phi_i\rangle$. In eq. (18.3) the first factor on the l.h.s. is a wave function diagram, while the second factor is just a matrix element. It is then clear that the second parenthesis in eq. (18.1) is simply $\langle j|U_V(0,-\infty)|i\rangle$, which is related to $|\chi_i^P\rangle$ by

$$|\chi_i^P\rangle = \sum_{j=1}^{d} |j\rangle\langle j|U_V(0,-\infty)|i\rangle \, . \tag{18.4}$$

Thus, we finally obtain from eqs. (16), (18.1) and (18.4)

$$U_V(0,-\infty)|\Phi_i\rangle = \sum_{j=1}^{d} U_{VQ}(0,-\infty)|\Phi_j\rangle\langle\Phi_j|U_V(0,-\infty)|\Phi_i\rangle \, , \tag{19}$$

where

$U_{VQ}(0,-\infty)|\Phi_j\rangle$ =

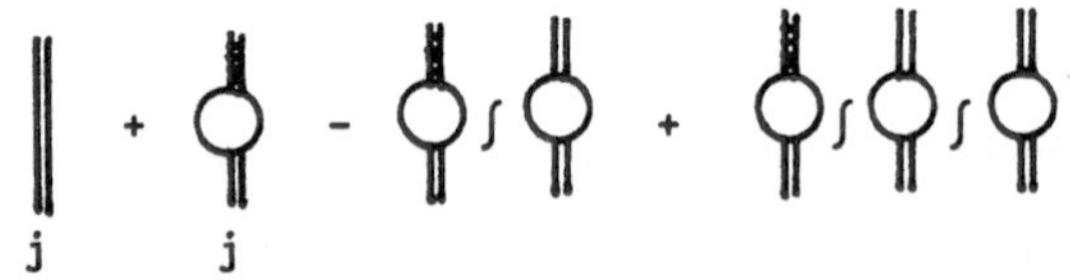

− ... (19.1)

and

$\langle j|U_V(0,-\infty)|i\rangle$ =

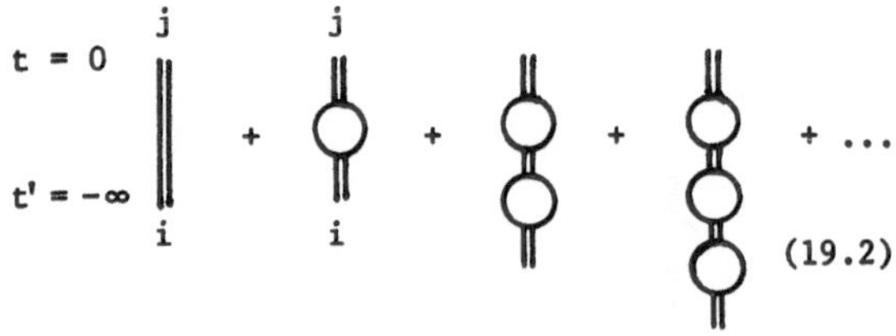

(19.2)

where clearly the first term vanishes if $i \neq j$. In eq. (19.1) all the terms are wave function diagrams, while the terms in eq. (19.2) are just matrix elements.

The final form of the decomposition theorem is then given by eqs. (12.2), (19) and (13), which for convenience are collected below:

$$U(0,-\infty)|\Phi_i\rangle = U_V(0,-\infty)|\Phi_i\rangle \times U(0,-\infty)|c\rangle \, , \tag{20}$$

$$U_V(0,-\infty)|\Phi_i\rangle = \sum_{j\le d} U_{VQ}(0,-\infty)|\Phi_j\rangle\langle\Phi_j|U_V(0,-\infty)|\Phi_i\rangle , \tag{20.1}$$

$$U(0,-\infty)|c\rangle = U_Q(0,-\infty)|c\rangle\langle c|U(0,-\infty)|c\rangle . \tag{20.2}$$

As noted above, eqs. (20.1-2) are closely similar to each other. The above theorem can be given the following pictorial representation

$U(0,-\infty)|\Phi_i\rangle = \langle U\rangle \times$  $\times \sum_{j\le d}$  $\times \langle\Phi_j|U_V(0,-\infty)|\Phi_i\rangle$, (21)

where $\langle U\rangle = \langle c|U(0,-\infty)|c\rangle$. The subscript V implies that all the vertices are linked to at least one valence line, while the index Q implies that except for the non-interacting term (delta function), all wave functions are passive at $t = 0$. Note that both $\langle U\rangle$ and $\langle\Phi_j|U_V|\Phi_i\rangle$ are divergent, since they contain terms with zero energy denominators. Thus, the theory will be designed so that these terms cancel away. On the other hand, each term in the expansion of $U_Q|c\rangle$ is finite, since all the vertices are linked to the time $t = 0$ via fermion lines, and thus there are no vanishing energy denominators. Let us now make sure that each term in the expansion of $U_{VQ}|\Phi_j\rangle$ is finite.

The formal structure of $U_{VQ}|\Phi_j\rangle$ is given by eq. (19.1). The first term has no interaction. Hence, the state j remains to be the state j at $t = 0$. All the other terms have at least one interaction, and at $t = 0$ they all terminate in passive states (which have at least one passive line). In order to study the detailed structure of these terms, we take j to be a state with two valence particles, i.e. $|\Phi_j\rangle = a_\alpha^\dagger a_\beta^\dagger|c\rangle$.

Then, a typical term in the non-folded $\hat{Q}$-box is

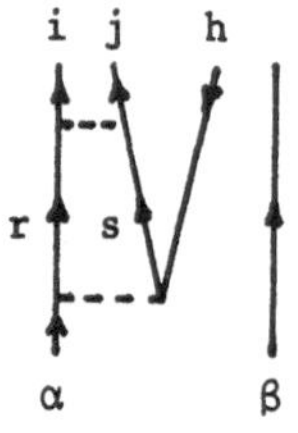

$$= a_i^{\dagger} a_j^{\dagger} a_h |c\rangle \left(\frac{1}{2}\right)^2 \frac{V_{ij,rs}\, V_{rs,\alpha h}}{(\varepsilon_\alpha - \varepsilon_i - \varepsilon_j + \varepsilon_h)(\varepsilon_\alpha - \varepsilon_r - \varepsilon_s + \varepsilon_h)} \; . \tag{22}$$

Diagram C of eq. (9) is a term belonging to $\hat{Q} \int \hat{Q}$ in eq. (19.1):

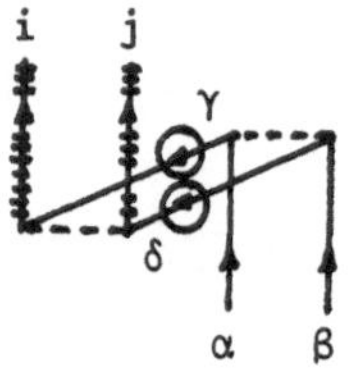

$$= a_i^{\dagger} a_j^{\dagger} |c\rangle \left(\frac{1}{2}\right)^2 \frac{V_{\gamma\delta,\alpha\beta}\, V_{ij,\gamma\delta}}{(\varepsilon_\alpha + \varepsilon_\beta - \varepsilon_i - \varepsilon_j)(\varepsilon_\gamma + \varepsilon_\delta - \varepsilon_i - \varepsilon_j)} \; . \tag{23}$$

As a third example, we mention that the twice-folded diagram of eq. (17) belongs to the $\hat{Q} \int \hat{Q} \int \hat{Q}$ term. From these examples it is clear that all the terms in $U_{VQ}|c\rangle$ are finite as long as the matrix elements of V are finite. (When V is singular, we shall use the G-matrix partial summation instead, as discussed in sect. 7.)

For future use it is convenient to write the decomposition theorem (20-20.2) in the form

$$U(0,-\infty)|\Phi_i\rangle = \sum_{j \le d} \Omega_V(0,-\infty)|\Phi_j\rangle\langle\Phi_j|U_V(0,-\infty)|\Phi_i\rangle\langle c|U(0,-\infty)|c\rangle \, , \tag{24}$$

where we have introduced

$$\Omega_V(0,-\infty)|\Phi_j\rangle = U_{VQ}(0,-\infty)|\Phi_j\rangle \times U_Q(0,-\infty)|c\rangle \, . \tag{24.1}$$

4. Derivation of the effective Hamiltonian

We are now prepared to derive a linked-diagram expansion of the effective Hamiltonian to be used in the model-space eigenvalue problem (4), where the model space is defined by the projection operator (3).

It is easily shown that by using the time-evolution operator in the complex-time limit we can construct, starting from a model-space state Φ_i, the lowest eigenstate Ψ_m of the true Hamiltonian H with $\langle\Phi_i|\Psi_m\rangle \neq 0$. The prescription for constructing $|\Psi_m\rangle$ is then†

$$\lim_{t'\to-\infty(\varepsilon)} \frac{U(0,t')|\Phi_i\rangle}{\langle\Phi_i|U(0,t')|\Phi_i\rangle} = \lim_{t'\to-\infty(\varepsilon)} \frac{\exp[iHt'/\hbar]\sum_n|\Psi_n\rangle\langle\Psi_n|\Phi_i\rangle}{\sum_\ell\langle\Phi_i|\exp[iHt'/\hbar]|\Psi_\ell\rangle\langle\Psi_\ell|\Phi_i\rangle}$$

$$= \frac{|\Psi_m\rangle}{\langle\Phi_i|\Psi_m\rangle} . \tag{25}$$

Here, only the lowest eigenstate Ψ_m survives because of the exponential damping factor. Note, however, that the above procedure is not unique, as different states Φ_i may lead to the same Ψ_m. To be more specific, let us apply the prescription of eq. (25) to the example of ^{18}O discussed in sect. 1. Suppose that we want to study the low-lying $J^\pi = 0^+$ states of ^{18}O, denoting the lowest one by $|0_1^+\rangle$. We expect these low-lying states to have non-zero overlaps with the model-space vectors $|\Phi_i\rangle$, i = 1, 2, 3, of eq. (2.2). Then, we have

$$\lim_{t'\to-\infty(\varepsilon)} \frac{U(0,t')|\Phi_i\rangle}{\langle\Phi_i|U(0,t')|\Phi_i\rangle} = \frac{|0_1^+\rangle}{\langle\Phi_i|0_1^+\rangle} , \quad i = 1, 2, 3 . \tag{25.1}$$

Thus, all the three model-space vectors $|\Phi_i\rangle$ will lead to the same true ground state $|0_1^+\rangle$.

---

† Here, we use the Schrödinger representation for the convenience of the proof. However, the results obtained hold equally well in the interaction picture.

The above approach is not suitable to our present purpose. The model-space equation (4) is supposed to give d solutions, d being the dimensionality of the model space. We shall need a scheme where we can construct more than one eigenstate of H by operating on different model-space wave functions. We would like to have a set of d parent states $|\rho_i\rangle$ in the model space such that there is a one-to-one correspondence between parent states and true eigenstates

$$\frac{U(0,-\infty)|\rho_\lambda\rangle}{\langle\rho_\lambda|U(0,-\infty)|\rho_\lambda\rangle} = \frac{|\Psi_\lambda\rangle}{\langle\rho_\lambda|\Psi_\lambda\rangle}, \qquad \lambda = 1, 2, \ldots, d. \tag{26}$$

Here it is understood that the time-evolution operator $U(0,-\infty)$ is taken in the complex-time limit. Since $|\Psi_\lambda\rangle$ is an eigenstate of H with eigenvalue E_λ, eq. (26) implies

$$H\frac{U(0,-\infty)|\rho_\lambda\rangle}{\langle\rho_\lambda|U(0,-\infty)|\rho_\lambda\rangle} = E_\lambda\frac{U(0,-\infty)|\rho_\lambda\rangle}{\langle\rho_\lambda|U(0,-\infty)|\rho_\lambda\rangle}, \qquad \lambda = 1, 2, \ldots, d, \tag{27}$$

which is a basic equation for the derivation of the P-space effective Hamiltonian H_{eff}. Two questions must be raised at this point:

(i) Can we actually construct such a set of parent states $|\rho_\lambda\rangle$?

(ii) Which eigenvalues of H will be reproduced by eq. (27)?

First, consider question (i). Assume that the <u>projections</u> $|P\Psi_\lambda\rangle$ of d eigenstates of H onto the model space are <u>known</u> and that they are <u>linearly independent</u>. Then, writing

$$|\rho_\lambda\rangle = \sum_{i=1}^{d} C_i^{(\lambda)}|\Phi_i\rangle, \qquad \lambda = 1, 2, \ldots, d, \tag{28}$$

where $|\Phi_i>$ denotes the P-space basis states, we can make $|\rho_\lambda>$ satisfy

$$<\rho_\lambda|\Psi_\mu> = <\rho_\lambda|P\Psi_\mu> = 0 , \quad \text{for } \lambda \neq \mu = 1, 2, \ldots , d. \quad (28.1)$$

Here, the first equality holds because $|\rho_\lambda>$ is entirely contained in the P-space. The second equality is true because we have assumed that the projections $P\Psi_\mu$ are linearly independent. Thus, the vectors ρ_λ can be explicitly constructed by way of the Schmidt orthogonalization procedure. Note that we have in general

$$|\rho_\lambda> \neq |P\Psi_\lambda> , \quad (28.2)$$

$$<P\Psi_\lambda|P\Psi_\mu> \neq 0 , \quad \lambda \neq \mu , \quad (28.3)$$

$$<\rho_\lambda|\rho_\mu> \neq 0 , \quad \lambda \neq \mu . \quad (28.4)$$

For simplicity, we choose

$$<\rho_\lambda|\rho_\lambda> = 1 , \quad \lambda = 1, 2, \ldots , d . \quad (28.5)$$

The parent states $|\rho_\lambda>$ should be regarded merely as a mathematical device in order to derive H_{eff}, since their construction assumes knowledge of $|P\Psi_\lambda>$ which is not available until H_{eff} is known. Thus, the final expression for H_{eff} should not depend on $|\rho_\lambda>$.

Once $|\rho_\lambda>$ satisfies eq. (28.1), we have the one-to-one correspondence (26) between $|\rho_\lambda>$ and $|\Psi_\lambda>$. This is because

$$U(t,t')|\rho_\lambda> = e^{-iH(t-t')/\hbar}|\rho_\lambda> = \sum_\mu e^{-iE_\mu(t-t')/\hbar}|\Psi_\mu><\Psi_\mu|\rho_\lambda>$$

$$= e^{-iE_\lambda(t-t')/\hbar}|\Psi_\lambda><\Psi_\lambda|\rho_\lambda> + \sum_{\mu>d} e^{-iE_\mu(t-t')/\hbar}|\Psi_\mu><\Psi_\mu|\rho_\lambda> , \quad (29)$$

where in the last step we have used eq. (28.1). Then, in the complex-time limit, the second term on the far r.h.s. of eq. (29) will vanish compared to the first. A similar argument holds for the denominator on the l.h.s. of eq. (26), and hence eq. (26) is true. Then, eq. (27) follows immediately and can be used to derive H_{eff}. Substituting eq. (28) into eq. (27) we obtain

$$\sum_{i=1}^{d} C_i^{(\lambda)} \frac{HU(0,-\infty)|\Phi_i>}{<\rho_\lambda|U(0,-\infty)|\rho_\lambda>} = \sum_{j=1}^{d} C_j^{(\lambda)} \frac{E_\lambda U(0,-\infty)|\Phi_j>}{<\rho_\lambda|U(0,-\infty)|\rho_\lambda>} . \tag{30}$$

By the decomposition theorem (24), eq. (30) becomes

$$\sum_{i,k=1}^{d} C_i^{(\lambda)} H\Omega_V(0,-\infty)|\Phi_k> \frac{<\Phi_k|U_V(0,-\infty)|\Phi_i>}{<\rho_\lambda|U(0,-\infty)|\rho_\lambda>} <U>$$

$$= \sum_{j,\ell=1}^{d} C_j^{(\lambda)} E_\lambda \Omega_V(0,-\infty)|\Phi_\ell> \frac{<\Phi_\ell|U_V(0,-\infty)|\Phi_j>}{<\rho_\lambda|U(0,-\infty)|\rho_\lambda>} <U> , \tag{30.1}$$

where $<U> \equiv <c|U(0,-\infty)|c>$. By defining

$$b_k^{(\lambda)} = \sum_{i=1}^{d} C_i^{(\lambda)} \frac{<\Phi_k|U_V(0,-\infty)|\Phi_i>}{<\rho_\lambda|U(0,-\infty)|\rho_\lambda>} <U> , \tag{30.2}$$

we can rewrite eq. (30.1) as

$$\sum_{k=1}^{d} H\Omega_V(0,-\infty)|\Phi_k> b_k^{(\lambda)} = \sum_{\ell=1}^{d} E_\lambda \Omega_V(0,-\infty)|\Phi_\ell> b_\ell^{(\lambda)} . \tag{30.3}$$

From the definition (24.1) of $\Omega_V|\Phi_i>$ and eq. (19.1) it follows that the only P-space component in $\Omega_V|\Phi_i>$ is $|\Phi_i>$, hence

$$<\Phi_m|\Omega_V(0,-\infty)|\Phi_\ell> \;=\; \delta_{m\ell} \,. \tag{30.4}$$

Thus, by multiplying eq. (30.3) by $<\Phi_m|$, we readily have

$$\sum_{k=1}^{d} <\Phi_m|H_{eff}|\Phi_k> b_k^{(\lambda)} \;=\; E_\lambda b_m^{(\lambda)} \,, \tag{31}$$

where

$$H_{eff} \;=\; H\Omega_V(0,-\infty) \,. \tag{31.1}$$

Because the states Φ_i, $i = 1, 2, \ldots, d$, are the P-space basis states, eq. (30) is a secular equation entirely contained in the P-space. Thus, it is of the form

$$PH_{eff}P\Psi_\lambda \;=\; E_\lambda P\Psi_\lambda \,, \tag{32}$$

provided that we can identify $b_k^{(\lambda)}$ with the projection of Ψ_λ onto the P-space. This can easily be done. By multiplying eq. (26) by $<\Phi_k|$ and applying eqs. (28) and (20), we have

$$\frac{<\Phi_k|\Psi_\lambda>}{<\rho_\lambda|\Psi_\lambda>} \;=\; \sum_{i=1}^{d} c_i^{(\lambda)} \frac{<\Phi_k|U_V(0,-\infty)|\Phi_i><c|U(0,-\infty)|c>}{<\rho_\lambda|U(0,-\infty)|\rho_\lambda>} \;=\; b_k^{(\lambda)} \,, \tag{33}$$

where the last step follows from eq. (30.2). Thus, $b_k^{(\lambda)}$ is proportional to the projection of the true eigenstate $|\Psi_\lambda>$ onto the P-space vector $|\Phi_k>$. Note that the only dependence of the model-space eigenvalue problem on the parent state $|\rho_\lambda>$ is via $b_k^{(\lambda)}$, as shown by eq. (33). Having shown that eq. (31) has the right formal structure, we are not interested in $|\rho_\lambda>$ any more and may just solve eq. (31) directly for $b_k^{(\lambda)}$. Then, for known $b_k^{(\lambda)}$, eq. (33) serves to ensure that a parent state $|\rho_\lambda>$ with the initially assumed properties does indeed exist.

We can express eqs. (31-33) on a more compact form. Defining the vector

$$|b_\lambda> = \sum_{k=1}^{d} b_k^{(\lambda)} |\Phi_k> , \tag{33.1}$$

we immediately have

$$PH_{eff}P|b_\lambda> = E_\lambda |b_\lambda> , \tag{34}$$

where

$$|b_\lambda> = \frac{P|\Psi_\lambda>}{<\rho_\lambda|\Psi_\lambda>} , \tag{34.1}$$

since $P = \sum_{k\leq d} |\Phi_k><\Phi_k|$. Hence, we have obtained the desired model-space secular equation, which is a matrix equation of dimension d (i.e. the dimension of the model space P). The solution of eq. (34) will give d eigenvalues of the true Hamiltonian H. In question (ii) above we asked which d eigenvalues of H would be reproduced by H_{eff}. According to the complex-time approach these are the <u>lowest</u> (i.e. most negative) d eigenvalues with eigenvectors Ψ_λ, $\lambda = 1, 2, \ldots , d$, satisfying $P\Psi_\lambda \neq 0$ (i.e. non-zero projection onto the P-space). This statement could not have been made in the adiabatic approach[34].

In practice, we may not obtain the lowest d states of H with non-zero P-space overlaps, since we are in general not able to compute H_{eff} exactly. We shall evaluate H_{eff} by summing diagrams in perturbation theory, and the convergence properties of this expansion remain to be established.

Before analysing H_{eff} of eq. (34) in detail, we observe that although

$$\langle\Psi_\lambda|\Psi_\mu\rangle = \delta_{\lambda\mu}, \tag{34.2}$$

eq. (34.1) implies that in general

$$\langle b_\lambda|b_\mu\rangle \neq \delta_{\lambda\mu}. \tag{34.3}$$

This is to be expected, since the projection of orthogonal vectors onto a smaller space does not in general lead to orthogonal projected vectors. Clearly, eq. (34.3) implies that H_{eff} is generally not Hermitian. In order to overcome the practical inconvenience caused by the non-orthogonality of the model-space eigenvectors $|b_\lambda\rangle$, we may define vectors $|\bar{b}_\lambda\rangle$ in the P-space which are biorthogonal to $|b_\lambda\rangle$, i.e.

$$\langle\bar{b}_\lambda|b_\mu\rangle = \delta_{\lambda\mu}. \tag{34.4}$$

Now, we proceed to evaluate the matrix elements of H_{eff} in the P-space. From eq. (31.1) we have

$$\langle\Phi_m|H_{eff}|\Phi_k\rangle = \langle\Phi_m|H\Omega_V(0,-\infty)|\Phi_k\rangle. \tag{35}$$

Recall from eq. (24.1) that $\Omega_V|\Phi_k\rangle$ consists of two factors

$$\Omega_V(0,-\infty)|\Phi_k\rangle = U_{VQ}(0,-\infty)|\Phi_k\rangle \times U_Q(0,-\infty)|c\rangle. \tag{35.1}$$

In the first factor all the interactions must be linked to at least one valence line, whereas in the second factor no valence lines are involved.

On the r.h.s. of eq. (35) we write $H = H_o + H_1$, where $H_o = T + U$ and $H_1 = V - U$. Consider first the contribution from H_o. Since Φ_m is an eigenstate of H_o and the only P-space component in $\Omega_V|\Phi_k\rangle$ is $|\Phi_k\rangle$ itself [see eq. (30.4)], we have

$$\langle\Phi_m|H_o\Omega_V(0,-\infty)|\Phi_k\rangle = \langle\Phi_m|H_o|\Phi_k\rangle = \langle\Phi_m|[H_o(V) + H_o(C)]|\Phi_k\rangle$$

$$= \delta_{mk}(W_V + W_C) . \qquad (35.2)$$

Here, H_o is split into a valence part $H_o(V)$ and a core part $H_o(C)$ with eigenvalues W_V and W_C, respectively. In the example of ^{18}O, W_V is the unperturbed energy of the two valence particles [e.g. $2\varepsilon(0d_{5/2})$ for $k = 1$, see eq. (2.2)] and W_C the unperturbed energy of the ^{16}O core.

Then, consider the contribution of H_1 to the matrix element (35)

$$\langle\Phi_m|H_1\Omega_V(0,-\infty)|\Phi_k\rangle = \langle\Phi_m|[H_1(V) + H_1(C)]\Omega_V(0,-\infty)|\Phi_k\rangle , \qquad (35.3)$$

where $H_1(C)$ gives rise to diagrams in which H_1 is not linked to any valence line at the time $t = 0$ and $H_1(V)$ to all the other diagrams. A diagram resulting from $H_1(C)$ is

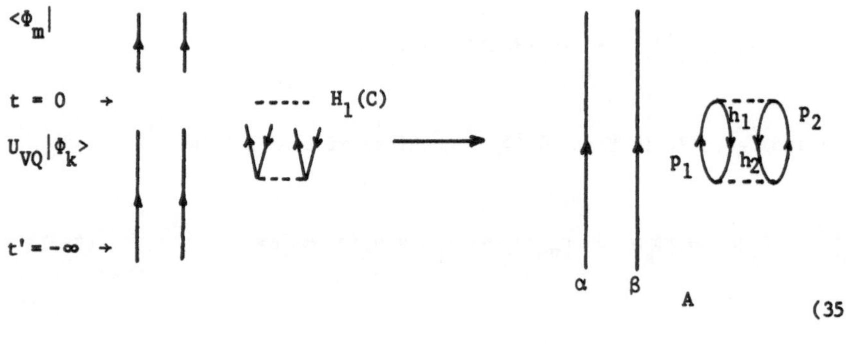

(35.4)

Clearly, diagram A is independent of α and β (except that this implies $|\Phi_m\rangle = |\Phi_k\rangle$). Thus, summing all the contributions from $H_1(C)$ we obtain

$$\langle\Phi_m|H_1(C)\Omega_V(0,-\infty)|\Phi_k\rangle =$$

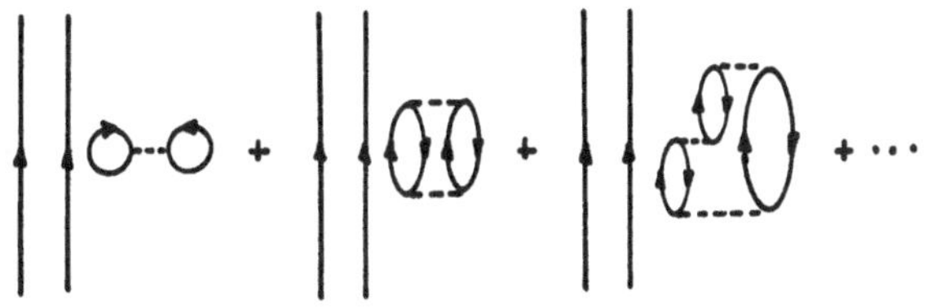

$$= \delta_{mk}\langle c|H_1U_Q(0,-\infty)|c\rangle = \delta_{mk}(E_C - W_C)\,, \qquad (35.5)$$

where E_C is the true core energy. The last step was made using the Goldstone theorem†. We then have for the matrix element (35)

$$\langle\Phi_m|H\Omega_V(0,-\infty)|\Phi_k\rangle = E_C\delta_{mk} + W_V\delta_{mk} + \langle\Phi_m|H_1(V)\Omega_V(0,-\infty)|\Phi_k\rangle\,. \qquad (35.6)$$

Using this in the secular equation (31), we obtain

$$\sum_{k=1}^{d}\langle\Phi_m|[H_o(V) + H_1(V)\Omega_V(0,-\infty)]|\Phi_k\rangle\, b_k^{(\lambda)} = (E_\lambda - E_C)b_m^{(\lambda)}\,. \qquad (36)$$

Thus, we have succeeded in separating out the core energy, as suggested by eq. (4.1). (In the example of ^{18}O considered above the eigenvalue $E_\lambda - E_C$ implies that the energy of ^{18}O is measured with respect to the ground-state energy of ^{16}O.) Furthermore, all the terms on the l.h.s. of eq. (36) are linked to at least one valence line. It is convenient to rewrite eq. (36) as

$$\sum_{k=1}^{d}\langle\Phi_m|[H_o(V) + \mathcal{V}_{eff}]|\Phi_k\rangle\, b_k^{(\lambda)} = (E_\lambda - E_C)b_m^{(\lambda)}\,, \qquad (36.1)$$

---

† Goldstone's theorem states that the energy shift $E_C - W_C$ of a non-degenerate system (e.g. the ground state of a closed-shell system) is equal to the sum of all the linked diagrams[37].

where

$$\langle\Phi_m|v_{\text{eff}}|\Phi_k\rangle = \langle\Phi_m|H_1(V)\Omega_V(0,-\infty)|\Phi_k\rangle \tag{36.2}$$

defines the so-called <u>effective interaction</u>. The valence-space secular equation (36.1) is completely analogous to the shell-model secular equation. Thus, eq. (36.2) may be considered as formally linking the shell-model effective interaction to the original nucleon-nucleon interaction.

Then, it remains to analyse v_{eff} and provide a prescription for calculating it. Recall from eq. (35.1)

$$\Omega_V|\Phi_k\rangle = U_{VQ}|\Phi_k\rangle \times U_Q|c\rangle$$

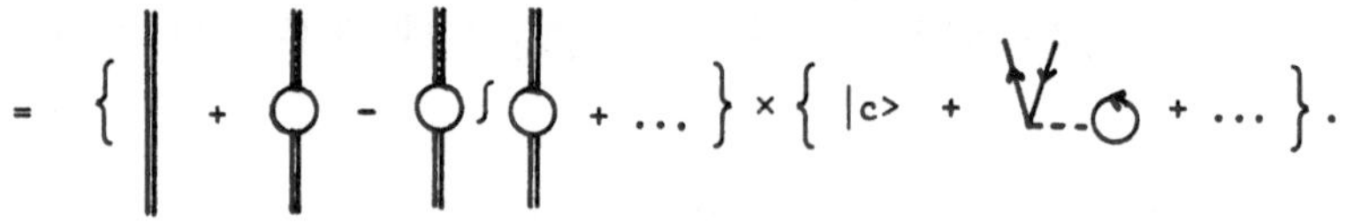

A B C D E (37)

The diagrammatic representation of the first factor follows from eq. (19.1) and of the second factor from fig. 7. Now, we can use eq. (37) to derive the diagrams contributing to the matrix element (36.2). The prescription is the following: In each term resulting from eq. (37) attach an interaction H_1 to at least one valence line at the time $t = 0$ and let the diagram terminate in the state Φ_m. Thus, if we take the P-space to consist of two valence lines, as in the example of ^{18}O, we obtain the following types of contributions to the matrix element (36.2):

(i) From $A \times D$ in eq. (37) we obtain

A × D ⟶ 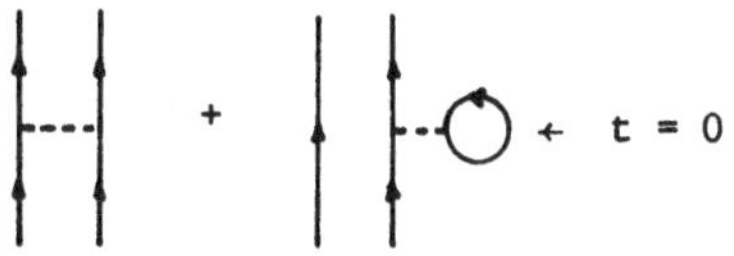 (37.1)

Note however that the first term on the r.h.s. of eq. (35.5) is not allowed, since the interaction is not connected to any valence line.

(ii) From B × D in eq. (37) we obtain terms like

B × D ⟶ 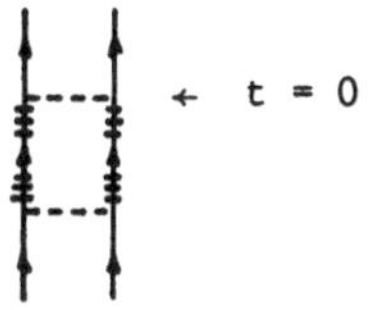 (37.2)

Note that this type of terms must have at least two H_1 vertices. Clearly, the diagrams in (i) and (ii) belong to the $\hat{Q}$-box  .

(iii) From C × D in eq. (37) we obtain terms like

C × D ⟶ 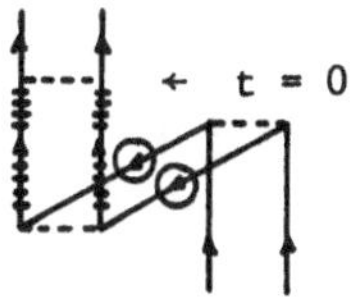 (37.3)

which is of the form  $\int$  . Here, we have distinguished between $\hat{Q}'$ and $\hat{Q}$. From eq. (37) it is obvious that $\hat{Q}'$ has at least two H_1 vertices, one from $H_1(V)$ and the other from $U_{VQ}|\Phi_k>$, while $\hat{Q}$ may have only the $H_1(V)$ vertex.

Thus, if only the term D is included in $U_Q|c>$, the matrix element of v_{eff} is given by

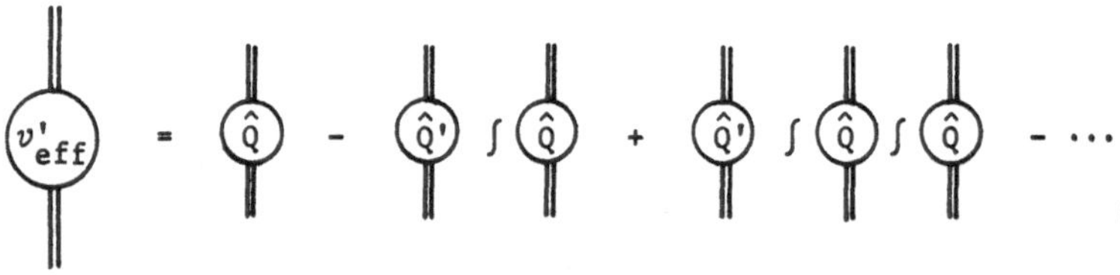

$$\left(v'_{eff}\right) = \left(\hat{Q}\right) - \left(\hat{Q}'\right)\int\left(\hat{Q}\right) + \left(\hat{Q}'\right)\int\left(\hat{Q}\right)\int\left(\hat{Q}\right) - \cdots \tag{37.4}$$

Here, $\hat{Q}$ is the sum of all irreducible diagrams with at least one vertex, and where all the vertices are linked to at least one valence line. Furthermore, $\hat{Q}'$ is obtained from $\hat{Q}$ by dropping the terms with fewer than two vertices. Thus, any <u>folded</u> diagram of v'_{eff} must have at least <u>three</u> interaction vertices.

Next, we return to eq. (37) and examine the contribution to v_{eff} from diagram E and the higher-order terms in $U_Q|c\rangle$. They will give rise to subtle differences between the $\hat{Q}'$- and $\hat{Q}$-boxes. The common feature of these contributions is that the core and valence pieces are joined together at $t = 0$ by the interaction $H_1(V)$. Examples are

(iv)   From $A \times E$ in eq. (37) we obtain

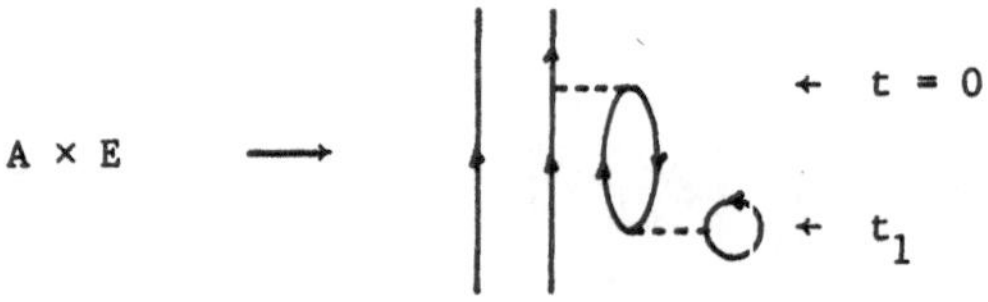

$$A \times E \quad \longrightarrow \tag{37.5}$$

Note that between $t = 0$ and $t = t_1$ there are only <u>active</u> lines in the valence part of the diagram.

(v)   From $B \times E$ in eq. (37) we obtain terms like

$$B \times E \quad \longrightarrow \tag{37.6}$$

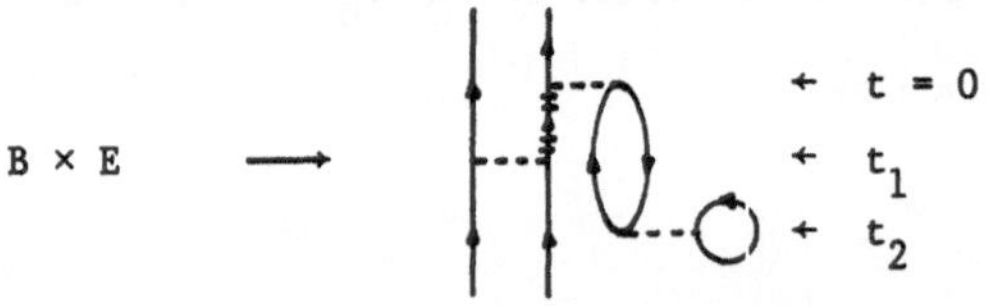

Note that before $t = 0$ there is at least one passive line in the valence part of the diagram. This is because diagram B terminates in a passive state.

(vi) From $C \times E$ in eq. (37) we obtain terms like

$C \times E \longrightarrow$ 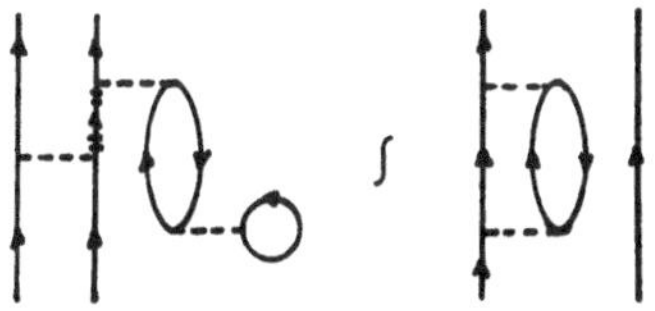 (37.7)

Note again that the valence part has at least one passive line prior to the interaction $H_1(V)$ at $t = 0$.

We shall call the diagrams in (iv)-(vi) last-moment core-insertion diagrams, since the core and valence parts are joined together at $t = 0$ by $H_1(V)$. Their general structure is of the types I-III shown in fig. 9. Types I and II are allowed in $\hat{Q}$, but only type II is allowed in $\hat{Q}'$. In $\hat{Q}'$ there must be at least two vertices H_1 attached to the valence lines. Type III clearly belongs to the folded diagrams. As shown in fig. 9 there is a fourth type (IV) of last-moment core-insertion diagrams. Such diagrams are only allowed in the $\hat{Q}$-boxes of the second, third, etc., folded terms of eq. (37.4), but not in the first, non-folded $\hat{Q}$, which we shall denote by $\hat{Q}_1$.

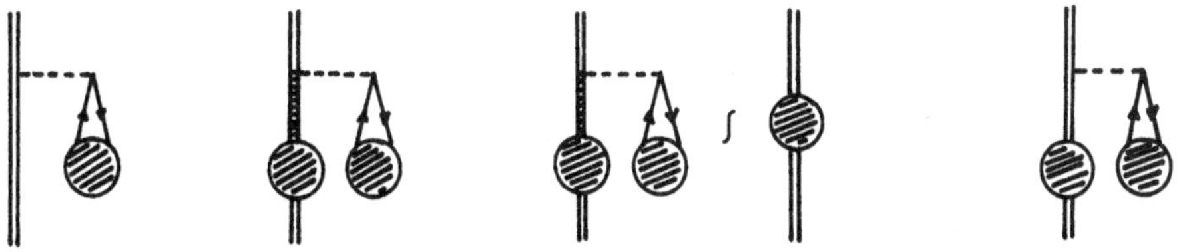

I ($\hat{Q}_1$, $\hat{Q}$)   II ($\hat{Q}_1$, $\hat{Q}'$, $\hat{Q}$)   III   IV ($\hat{Q}$)

Fig. 9. Various types of last-moment core-insertions. For details, see the text.

Thus, when all the contributions from $U_Q|c\rangle$ are included, v_{eff} has the following structure†

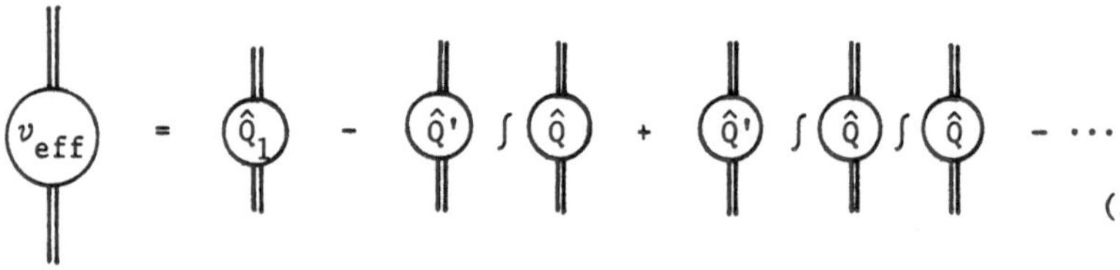

(38a)

$$= \hat{Q}_1 - \hat{Q}'\int\hat{Q} + \hat{Q}'\int\hat{Q}\int\hat{Q} - \cdots \qquad (38b)$$

Here, the different $\hat{Q}$-boxes $\hat{Q}_1$, $\hat{Q}'$ and $\hat{Q}$ are all composed of irreducible diagrams in which the vertices are linked to at least one valence line, $\hat{Q}'$ being at least second order in H_1 , while $\hat{Q}_1$ and $\hat{Q}$ both start with first order terms. Furthermore, there are the following differences in the above $\hat{Q}$-boxes as regards the inclusion of last-moment core-insertions (see fig. 9): In $\hat{Q}_1$ only types I and II are allowed, in $\hat{Q}'$ only type II is allowed, and in $\hat{Q}$ types I, II and IV are allowed.

We give an example to illustrate the above points. Consider the low-order $\hat{Q}$-box diagrams with two valence lines shown in fig. 10. In all the diagrams the top vertex is at $t = 0$. Among the diagrams in fig. 10, only diagram 8 (which is of type IV in fig. 9) is not allowed in $\hat{Q}_1$. Diagrams 1, 2, 6 and 8 are not allowed in $\hat{Q}'$, 1 and 2 because they are first order, and 6 and 8 because there are no passive lines in the valence parts.

For future application it will be convenient to distinguish between <u>valence-connected</u> and <u>valence-disconnected</u> diagrams. In a valence-connected diagram all the linked pieces are linked together into a single linked diagram. On the other hand, a valence-disconnected diagram consists of more than one linked piece. Diagram 5 in fig. 10 is thus

† This equation can be cast into a simpler form when one employs generalized time ordering for the core insertions, as will be shown in eq. (49).

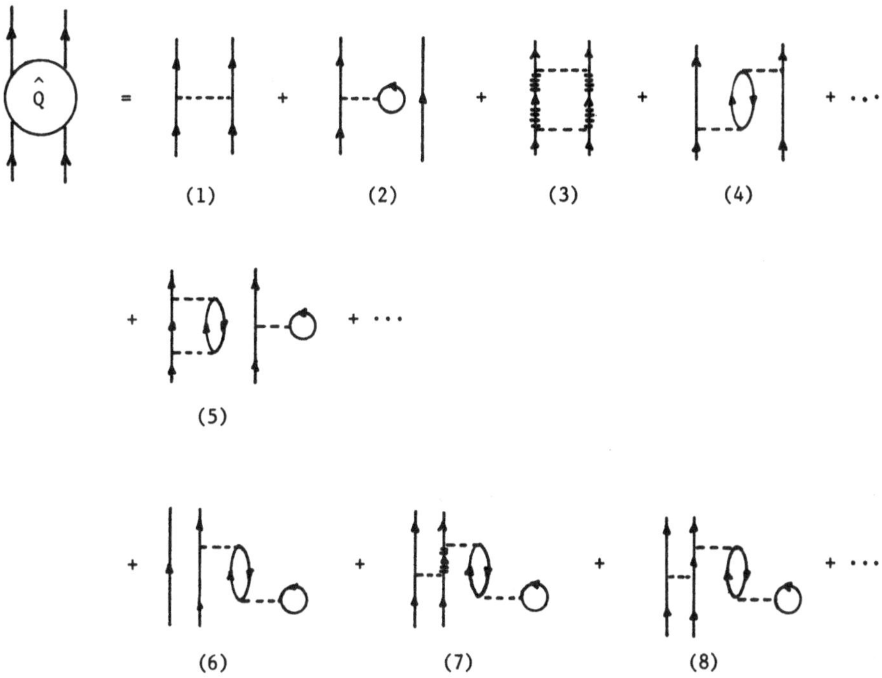

Fig. 10. $\hat{Q}$-box diagrams with two valence lines.

disconnected, but 2 is connected (recall that a linked piece must have at least one interaction). All the other diagrams in fig. 10 are clearly connected. Then, it is evident that v_{eff} consists of both valence-connected and valence-disconnected diagrams.

Let us summarize the developments thus far. We have reduced the the exact many-body problem (1) to the model-space eigenvalue problem (36.1-2). The model-space effective Hamiltonian is given as $H_o(V) + v_{eff}$, where $H_o(V)$ is known and v_{eff} is given algebraically by eq. (36.2) and diagramatically by eq. (38). All the terms in eq. (38) involve time integrals. This is obviously because our starting point for the derivation of v_{eff} is an analysis of the ratio [see eq. (30)]

$$\frac{\langle\Phi_m|HU(0,-\infty)|\Phi_k\rangle}{\langle\rho_\lambda|U(0,-\infty)|\rho_\lambda\rangle} \tag{39}$$

using time-dependent perturbation theory. In fact, v_{eff} is just the well-behaved part of $\langle\Phi_m|HU(0,-\infty)|\Phi_k\rangle$; the divergent terms in $\langle\Phi_m|HU(0,-\infty)|\Phi_k\rangle$ have been eliminated by the present formulation.

Consider again our example with two valence nucleons, corresponding to taking

$$|\Phi_m\rangle = a_\alpha^\dagger a_\beta^\dagger|c\rangle \quad \text{and} \quad |\Phi_k\rangle = a_\gamma^\dagger a_\delta^\dagger|c\rangle . \tag{39.1}$$

Then, the time integrals in the matrix element $\langle\alpha\beta|v_{eff}|\gamma\delta\rangle$ arise from

$$\lim_{t'\to-\infty(\varepsilon)} \langle c|a_\beta a_\alpha H_1(t=0)U(0,t')a_\gamma^\dagger a_\delta^\dagger|c\rangle . \tag{39.2}$$

To evaluate v_{eff} is just to evaluate the linked (and non-divergent) diagrams originating from the matrix element (39.2), in close similarity to the calculation of the energy shift in the Goldstone theorem[37].

To compute the matrix element (39.2), we must perform (i) all contractions and (ii) all time integrals. In doing so, we shall need the diagram rules for v_{eff}. These will be studied in detail in sect. 6. For the present, we shall just evaluate a few typical diagrams in v_{eff}, in order to familiarize ourselves with the procedure.

Consider the four diagrams A-D in fig. 11. Clearly, diagram A belongs to the terms $\hat{Q}_1$ and $\hat{Q}$ from $\langle\alpha\beta|v_{eff}|\gamma\delta\rangle$, while diagram B belongs to $\hat{Q}_1$, $\hat{Q}'$ and $\hat{Q}$. From expression (39.2) it is clear that diagram A is just

$$A = \frac{1}{2} V_{\alpha\beta,\gamma\delta} . \tag{39.3}$$

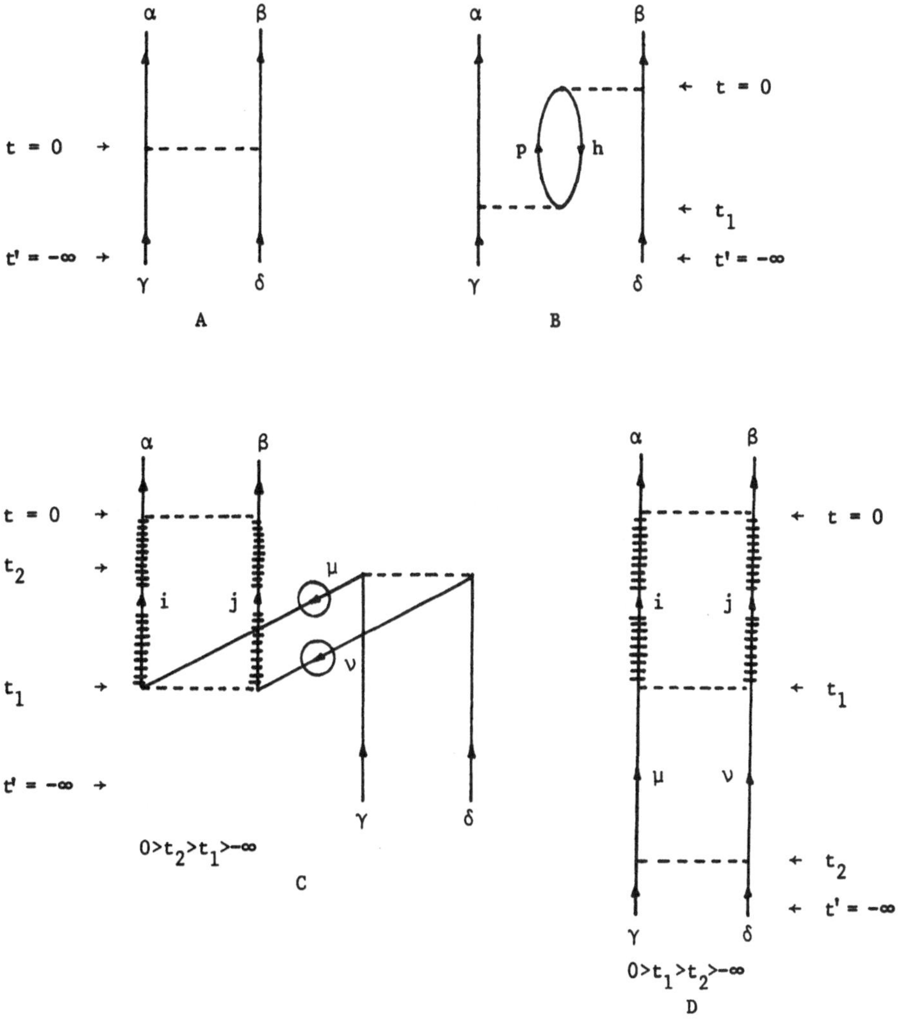

Fig. 11. Diagrams evaluated in eqs. (39.3-9).

Similarly, diagram B is

$$B = (-)^{n_\ell+n_h} \frac{1}{4} \sum_{p,h} V_{h\beta,p\delta} V_{\alpha p,\gamma h}$$

$$\times \lim_{t'\to -\infty(\varepsilon)} (\frac{-i}{\hbar}) \int_{t'}^{0} dt_1\, e^{-i(\varepsilon_\gamma+\varepsilon_\delta-\varepsilon_\alpha-\varepsilon_\delta-\varepsilon_p+\varepsilon_h)t_1/\hbar}$$

$$= \frac{1}{4} \sum_{p,h} \frac{V_{h\beta,p\delta} V_{\alpha p,\gamma h}}{\varepsilon_\gamma+\varepsilon_\delta-(\varepsilon_\alpha+\varepsilon_\delta+\varepsilon_p-\varepsilon_h)} , \qquad (39.4)$$

where we used $n_\ell = 1$ (one closed loop) and $n_h = 1$ (one hole line). The factor $(-)^{n_\ell+n_h}$ arises from the contractions of the fermion operators in the matrix element (39.2). We shall discuss the diagrammatic rules for $\mathcal{V}_{eff}$ in detail in sect. 6.

Diagram C in fig. 11 belongs to the term $-\hat{Q}'\int\hat{Q}$ in eq. (38). It is helpful to consider diagram D as well, which belongs to the matrix element (39.2). In both diagrams the last interaction is at $t = 0$. Furthermore, both the time-independent factors and the time-integrands of these diagrams are identical. Only the integration intervals are different. Thus, we can express diagrams C and D as

$$\left.\begin{matrix} C \\ D \end{matrix}\right\} = \frac{1}{8} \sum_{i,j\in Q} \sum_{\mu,\nu\in P} V_{\alpha\beta,ij} V_{ij,\mu\nu} V_{\mu\nu,\gamma\delta} \times \left\{\begin{matrix} I(C) \\ I(D) \end{matrix}\right. , \qquad (39.5)$$

where the time integrals are

$$I(C) = \lim_{t'\to -\infty(\varepsilon)} \int_{t'}^{0} dt_2 \int_{t'}^{t_2} dt_1\, I(t_1,t_2) \qquad (39.6)$$

and

$$I(D) = \lim_{t'\to -\infty(\varepsilon)} \int_{t'}^{0} dt_1 \int_{t'}^{t_1} dt_2\, I(t_1,t_2) , \qquad (39.7)$$

the integrand being

$$I(t_1,t_2) = (\frac{-i}{\hbar})^2 e^{-i(\varepsilon_\mu+\varepsilon_\nu-\varepsilon_i-\varepsilon_j)t_1/\hbar} e^{-i(\varepsilon_\gamma+\varepsilon_\delta-\varepsilon_\mu-\varepsilon_\nu)t_2/\hbar} . \qquad (39.8)$$

Clearly, I(D) is divergent if $\varepsilon_\gamma+\varepsilon_\delta = \varepsilon_\mu+\varepsilon_\nu$, i.e. for a degenerate P-space. Then, by factorizing diagram D as we did in the derivation of v_{eff}, we obtain diagram C, which is finite and gives

$$I(C) = \frac{1}{(\varepsilon_\gamma+\varepsilon_\delta-\varepsilon_i-\varepsilon_j)(\varepsilon_\mu+\varepsilon_\nu-\varepsilon_i-\varepsilon_j)} . \qquad (39.9)$$

Thus, we have made the following important observation: The matrix element $\langle\alpha\beta|v_{eff}|\gamma\delta\rangle$ is the sum of all the valence-linked (i.e. H_1 must be linked to at least one valence line) diagrams of $\langle c|a_\beta a_\alpha H_1(t=0)U(0,-\infty)a_\gamma^\dagger a_\delta^\dagger|c\rangle$ with all intermediate active states folded. Note that diagrams with active intermediate states may be divergent, but will become finite when the active intermediate states are folded.

In table 1 we show schematically how various terms in the matrix element $\langle c|a_\beta a_\alpha H_1(t=0)U(0,-\infty)a_\gamma^\dagger a_\delta^\dagger|c\rangle$, as grouped according to the number of active intermediate states, contribute to $\langle\alpha\beta|v_{eff}|\gamma\delta\rangle$. The one-$\hat{Q}$-box term (I) contributes directly to v_{eff} since it contains no active intermediate states. However, the two- and three-$\hat{Q}$-box terms (II and III) must be folded once and twice, respectively, before their contributions to v_{eff} can be obtained. This is because they contain one and two active intermediate states, respectively. It is further noted from table 1 that the leading $\hat{Q}$-box (i.e. the one to the left) has its top interaction at $t = 0$. All other interactions take place prior to $t = 0$. Note also that we have ignored the differences between the various $\hat{Q}$-boxes shown in eq. (38).

Table 1. Various contributions from $H_1(t=0)U(0,-\infty)$ to v_{eff}.

	Term in $\langle c\vert a_\beta a_\alpha H_1(t=0)U(0,-\infty)a_\gamma^\dagger a_\delta^\dagger\vert c\rangle_V$	Contribution to $\langle c\vert a_\beta a_\alpha v_{eff} a_\gamma^\dagger a_\delta^\dagger\vert c\rangle$
(I)	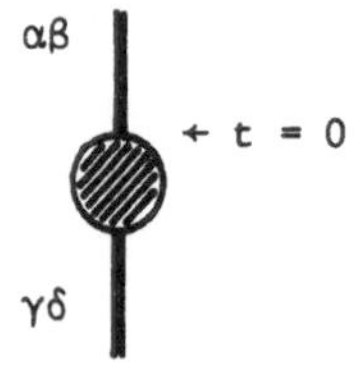	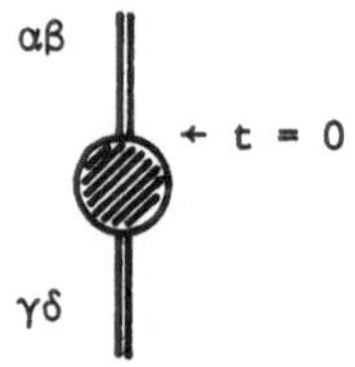
(II)	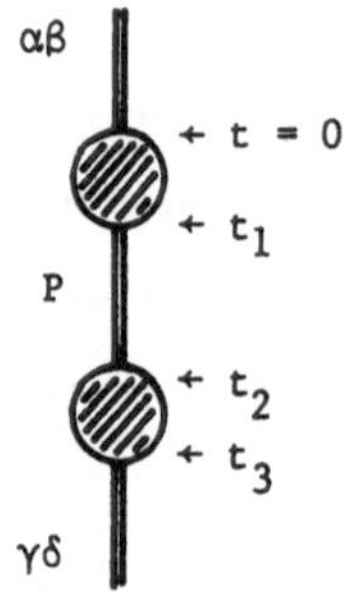 $0 > t_1 > t_2 > t_3 > -\infty$	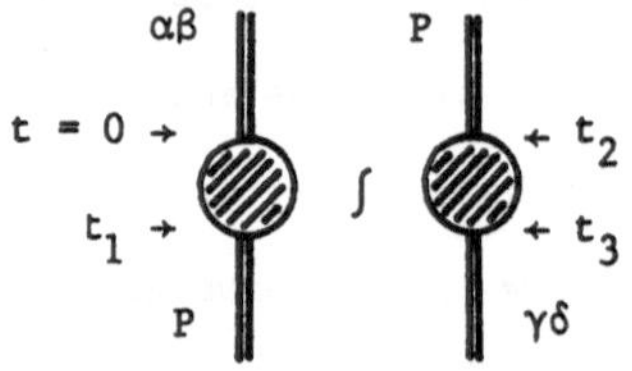 $0 > t_1 > -\infty, \quad 0 > t_2 > t_3 > -\infty$ $t_2 > t_1$
(III)	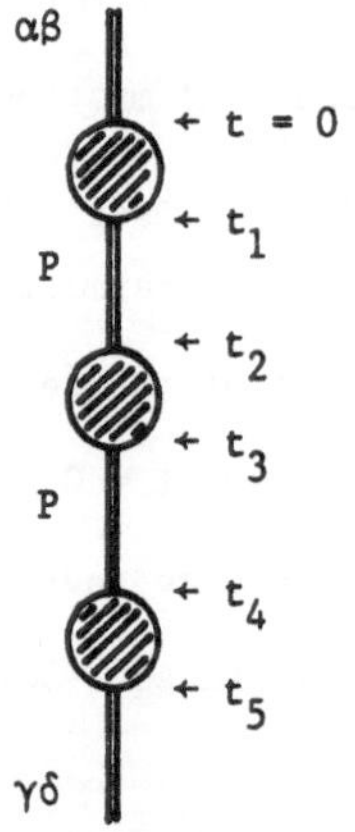 $0 > t_1 > t_2 > t_3 > t_4 > t_5 > -\infty$	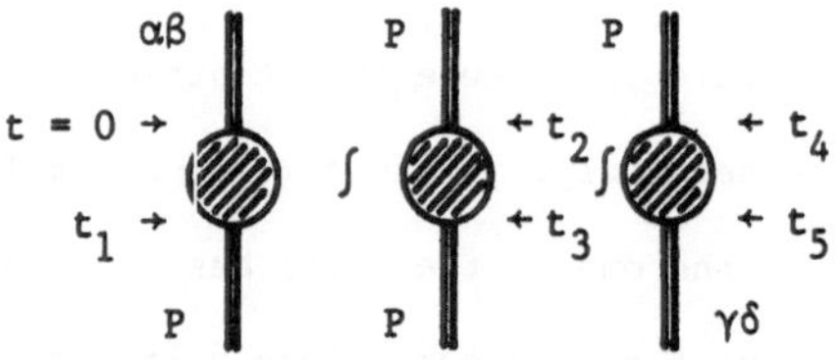 $0 > t_1 > -\infty, \quad 0 > t_2 > t_3 > -\infty, \quad 0 > t_4 > t_5 > -\infty$ $t_2 > t_1, \quad t_4 > t_3$

P indicates summation over all active states. It is understood that the folded terms carry an overall minus sign if the number of folds is odd. Note also that folding requires that the first $\hat{Q}$-box contains at least two H_1 interactions.

5. Properties of the folded-diagram expansion of the effective interaction

In order to understand the meaning of the effective interaction derived in the previous section, we shall discuss some of its properties in detail in this section. The following subjects will be considered:

(1) Cancellation of disconnected diagrams

(2) On-energy-shell core insertions

(3) Many-body effective interactions

(4) Folded diagrams and energy derivatives of $\hat{Q}$-boxes

(5) Connection with the Goldstone (non-degenerate) expansion.

5.1. Cancellation of disconnected diagrams. Consider the case with two valence particles discussed in the previous section. For this case, all the diagrams of v_{eff} have two outgoing and two incoming valence lines. As before, we use the symbols α, β, γ and δ to label these valence lines.

In the previous section we introduced the distinction between connected (i.e. valence-connected) and disconnected (i.e. valence-disconnected) diagrams. A disconnected diagram of v_{eff} has at least two separate linked pieces, each of which has at least one H_1 vertex. To make sure that we understand this definition, we shall give a simple example. In fig. 12 we show a few diagrams belonging to $\hat{Q}_1$ (A1-A3) and to $-\hat{Q}'\int\hat{Q}$ (A4-A5). Among the non-folded diagrams, A1 and A2 are connected, while A3 is disconnected. The folded diagram A4 is connected. For a folded diagram to be connected, it is sufficient that it be connected before or after folding. Thus, diagram A5 is clearly disconnected.

We shall now show that the two disconnected diagrams A3 and A5 exactly cancel each other. To see this, it is helpful to realize that a non-interacting fermion-line propagator can be folded any number of times. For example, we have

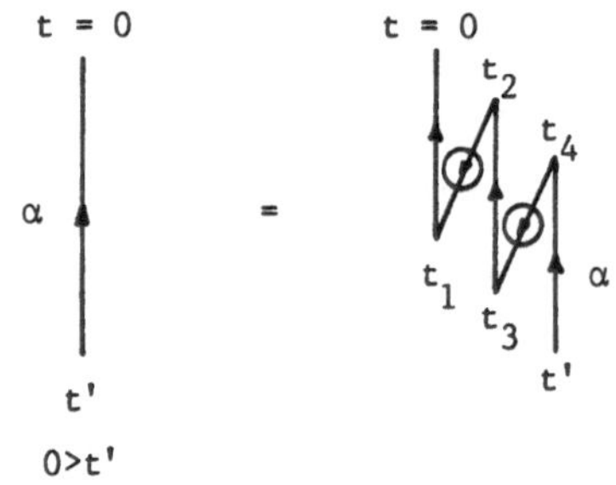

$0>t_1,\ t_2>t_3,\ t_4>t'$
$t_2>t_1,\ t_4>t_3$ (40)

This can be seen immediately by writing dcwn the propagators on both sides:

$$\text{LHS} = \exp\{-\tfrac{i}{\hbar}(0-t')\varepsilon_\alpha\}$$

$$= \exp\{-\tfrac{i}{\hbar}[(0-t_1) + (t_1-t_2) + (t_2-t_3) + (t_3-t_4) + (t_4-t')]\varepsilon_\alpha\} = \text{RHS}\ . \quad (40.1)$$

Thus, we can just stretch out non-interacting folded lines, and hence we see that A5 = -A3 , keeping in mind the minus sign of $-\hat{Q}'\int\hat{Q}$.

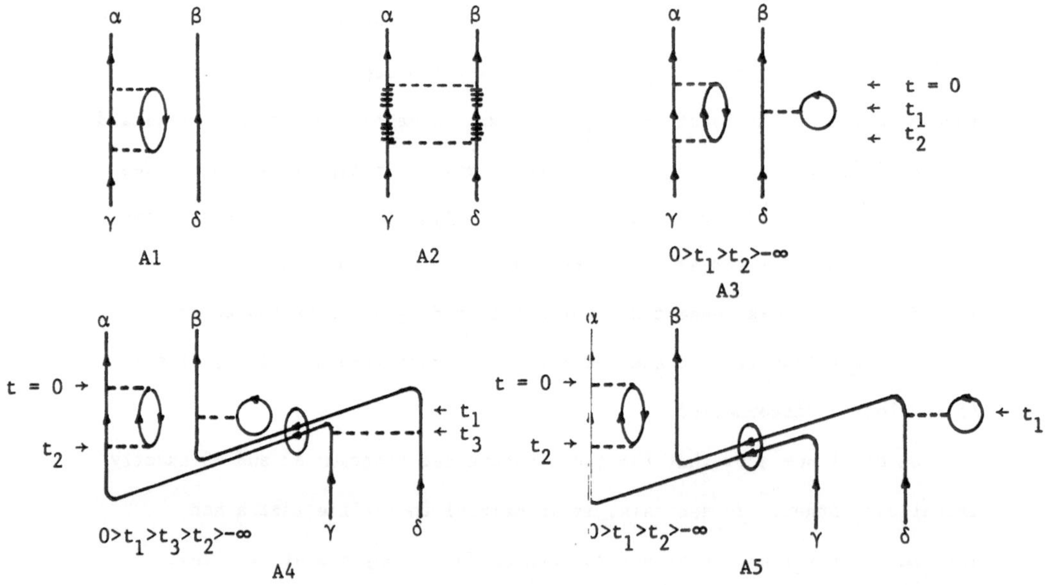

Fig. 12. Various diagrams belonging to $\hat{Q}_1$ (A1-A3) and to $-\hat{Q}'\int\hat{Q}$ (A4-A5). Diagrams A1-A2 and A4 are valence-connected, while A3 and A5 are valence-disconnected. It is understood that A4-A5 carry the minus sign of $-\hat{Q}'\int\hat{Q}$.

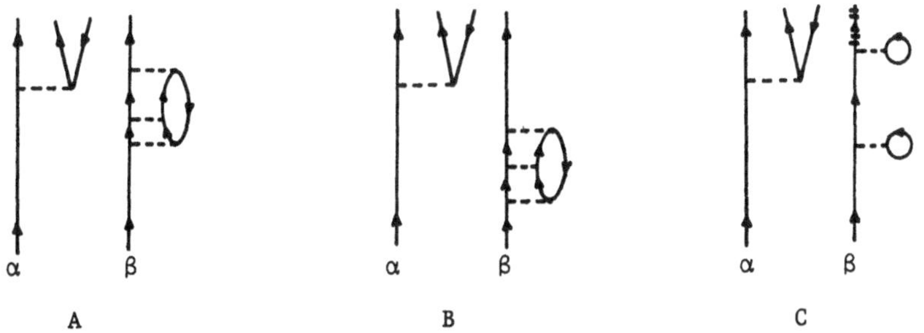

Fig. 13. Diagrams contained in $|\chi_i^Q\rangle$.

In fact, the cancellation of disconnected diagrams can easily be shown for many simple cases, just as in the example considered above. In the general case, it becomes difficult to prove the cancellation of disconnected diagrams by explicitly writing out the mutually cancelling terms. It is, however, possible to prove this cancellation by making use of the concept of generalized time ordering. The proof is in fact quite simple, as we shall now show.

Let us go back for a while to our discussion of the decomposition theorem [see eqs. (16-21)], where we wanted to extract a factor $|\chi_i^P\rangle$ from the wave function $|\chi_i^Q\rangle$. Since $|\chi_i^Q\rangle$ must be in a passive state at the time $t = 0$, the three terms shown in fig. 13 are all allowed in $|\chi_i^Q\rangle$. Each of these diagrams is composed of two disconnected branches, labelled by α and β. Clearly, their contributions to v_{eff} can be of the following types

A $\longrightarrow$ 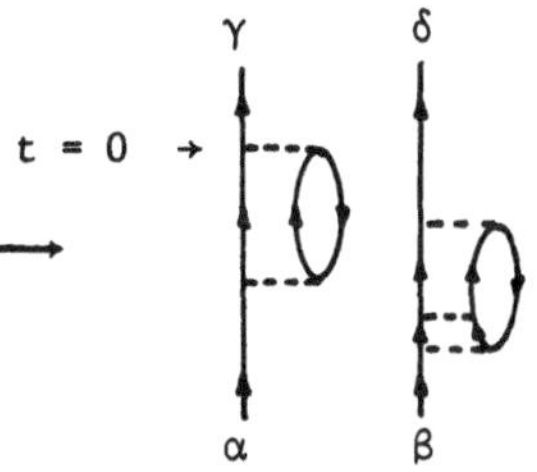

(41a)

B

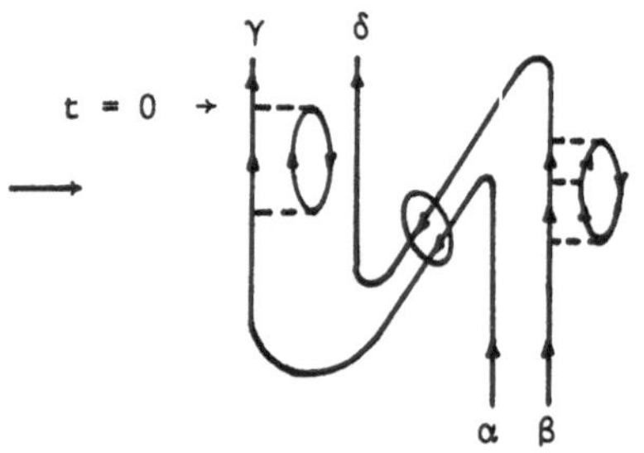

(41b)

C

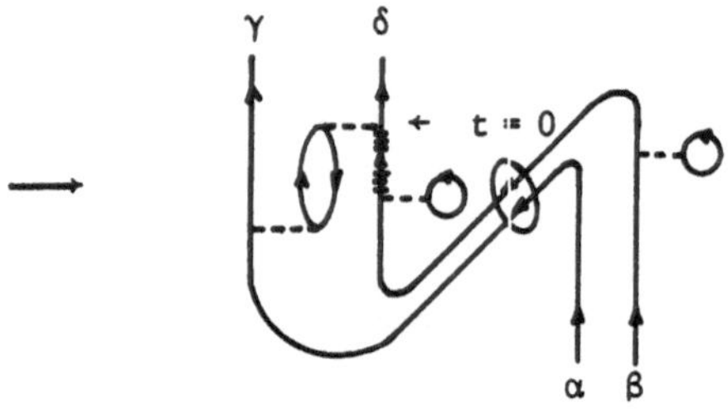

(41c)

As the matrix elements of v_{eff} are always taken between two active states (i.e. γ and δ are both restricted to be active), the terms in $|\chi_i^Q\rangle$ of type C can only produce connected diagrams of v_{eff}. For this type of terms both α and β terminate in a passive state. Since γ and δ are both active, the resulting v_{eff} is identically zero if the interaction at $t = 0$ is attached entirely to a single branch. Hence, in our study of the disconnected diagrams of v_{eff} we need not consider the terms in $|\chi_i^Q\rangle$ of type C.

Clearly, the terms in $|\chi_i^Q\rangle$ of types A and B can produce disconnected diagrams of v_{eff}, as shown by eqs. (41a-b). There are different schemes to factorize diagrams of the types A and B. The scheme used in sect. 3 treated them in the following way:

A 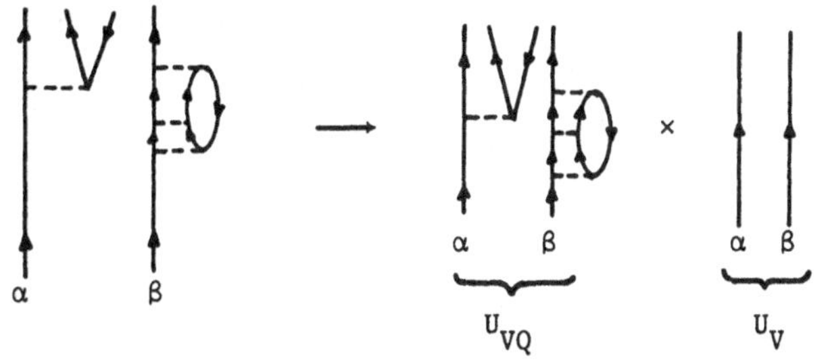

(42a)

B 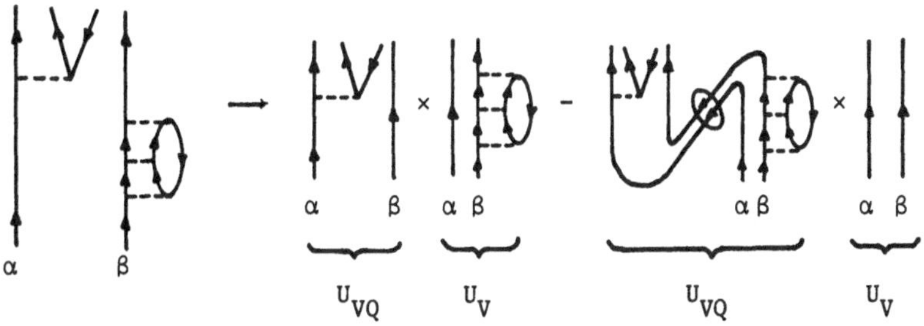

(42b)

Here, U_{VQ} and U_V denote, respectively, the factors $U_{VQ}|\Phi_j\rangle$ and $\langle\Phi_j|U_V|\Phi_i\rangle$ of eq. (19). To show the cancellation of disconnected diagrams in $\mathcal{V}_{eff}$, it would be sufficient to show that U_{VQ} does not contain any disconnected part which terminates in an active state. The U_{VQ} term of eq. (42a) and the folded U_{VQ} term of eq. (42b) are of this nature. In fact, they give rise to the two disconnected diagrams of $\mathcal{V}_{eff}$ shown in eqs. (41a) and (41b), respectively. These two terms cancel. For higher order diagrams, however, it is virtually impossible to prove the cancellation of disconnected diagrams term by term.

We therefore consider a more powerful scheme of factorization, where we make use of the concept of generalized time ordering (g.t.o.). We sum up a set of diagrams with some common properties, such as

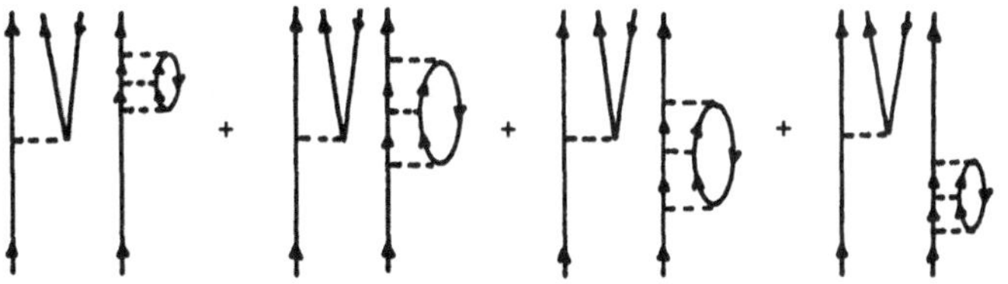

$\equiv A_{gto}$. (43)

Note that these diagrams are identical to each other except for different relative time ordering of the vertices. In fact, the third and fourth terms are, respectively, the diagrams A and B of fig. 13. By writing out the time integrals of the four terms in eq. (43), we readily obtain

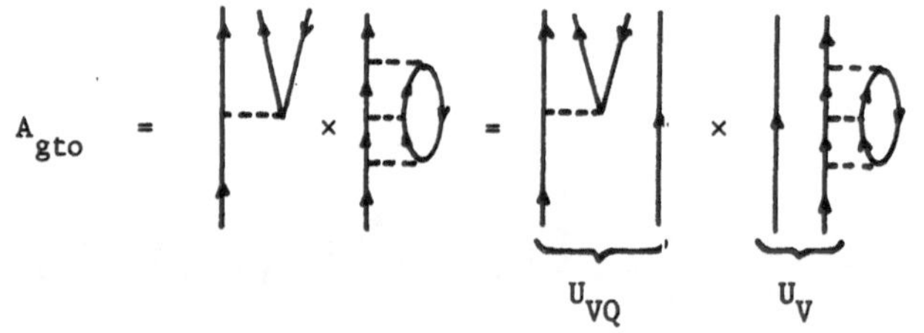

(44)

where we note that the g.t.o. factorization leads to a U_{VQ} term which is connected† and hence will not produce any disconnected diagrams of v_{eff}. This can be stated in a different way. Each of the four terms of $|\chi_i^Q\rangle$ shown in eq. (43) produce disconnected diagrams of v_{eff}. But these exactly cancel among themselves.

The above g.t.o. factorization of the disconnected terms contained in $|\chi_i^Q\rangle$ can readily be generalized. Such an example is given in fig. 14.

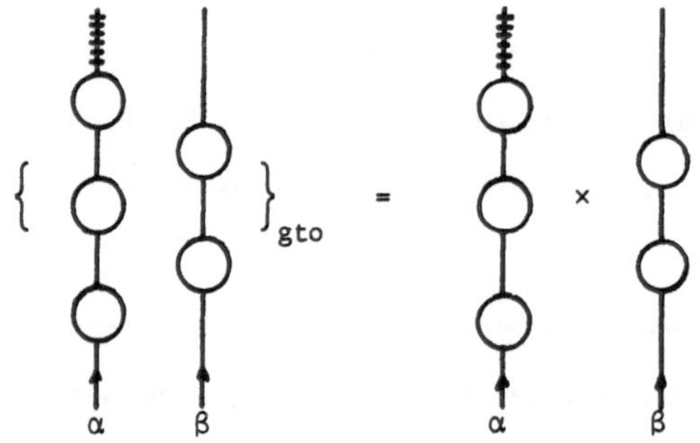

Fig. 14. A g.t.o. factorization of a general disconnected term in $|\chi_i^Q\rangle$.

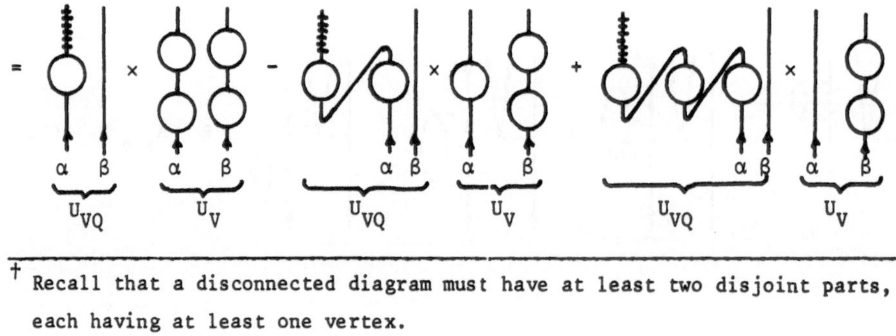

† Recall that a disconnected diagram must have at least two disjoint parts, each having at least one vertex.

We first sum up all the $\alpha - \beta$ disconnected terms of identical state labels but of different relative time ordering of the vertices. This enables us to write these terms as a product of two independent factors, as shown in fig. 14. Then the α-part is factorized by way of the folding procedure. As shown by the last equation of fig. 14, the resulting U_{VQ} terms are all connected. Hence, we have shown that v_{eff} does not contain any disconnected diagrams.

It should be pointed out that there are individual non-zero disconnected diagrams in v_{eff}. But these vanish when added up using g.t.o. Thus, as shown in fig. 15, one generally has to include diagrams from different folds to ensure cancellation. On the other hand, the disconnected diagrams cancel in the same order of the interaction. Hence, if we are calculating v_{eff} by summing diagrams order by order, one may leave out all the disconnected diagrams from the beginning. However, one does not of course do anything wrong by including the disconnected diagrams and letting the cancellation take care of itself. The above result is very similar to the cancellation of unlinked diagrams in the Goldstone expansion. There, unlinked terms are present, but are made to cancel exactly with each other in each order by introducing Pauli-violating diagrams.

Our analysis of the structure of U_{VQ} with g.t.o. has shown that v_{eff} does not contain any disconnected diagrams. It may be of interest to note that from this analysis we can show that a certain type of connected diagrams is not allowed in v_{eff}. Consider the example shown in fig. 16. This diagram arises from a U_{VQ} term composed of two disconnected parts, the δ-part being in an active state just prior to the time $t = 0$. We have shown that under g.t.o. U_{VQ} does not contain any such disconnected term, and hence connected diagrams of this type are not allowed in v_{eff}. (Allowed disconnected terms in U_{VQ} must have all their individual parts in passive states just before the time $t = 0$.) In other words, the diagram of fig. 16 will be cancelled by the folded diagram obtained by folding the

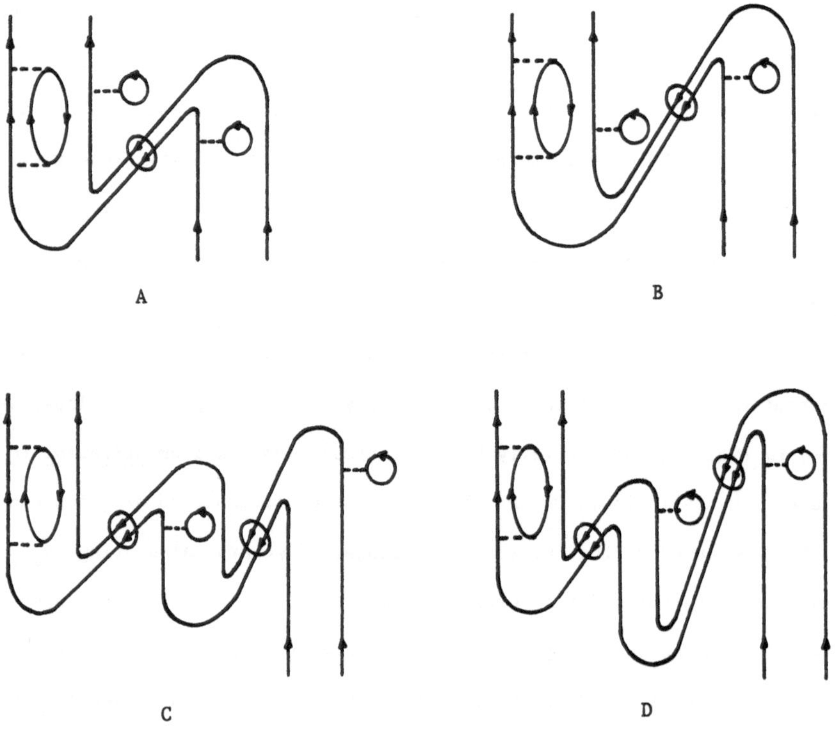

Fig. 15. Cancellation of disconnected diagrams in v_{eff}. Note that each individual diagram does not vanish in general, but the sum of all the four diagrams is zero. In fact, the diagrams are identical in all their factors, but differ in sign due to the different number of folds.

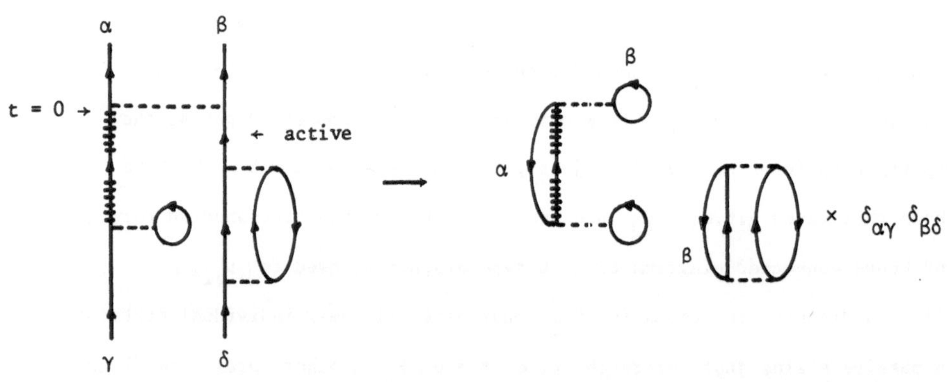

Fig. 16. Diagram not allowed in v_{eff}. Note that the corresponding Goldstone diagram referring to the vacuum state $|c'\rangle = a_\alpha^\dagger a_\beta^\dagger |c\rangle$ is unlinked, as shown on the right.

active line prior to the interaction at $t = 0$. In fact, the corresponding Goldstone diagram is unlinked, as shown on the r.h.s. of fig. 16.

A brief summary may be helpful. We have shown that v_{eff} does not contain any <u>disconnected</u> diagrams as they exactly cancel each other. Further certain connected diagrams cancel among themselves. Their common structure is that they are composed of two disjoint pieces linked together only by the $t = 0$ interaction, one piece being in an active state prior to this interaction.

<u>5.2. On-energy-shell core insertions.</u> Recall the expansion (38b) of v_{eff} in terms of $\hat{Q}$-boxes:

$$v_{eff} = \hat{Q}_1 - \hat{Q}'\int\hat{Q} + \hat{Q}'\int\hat{Q}\int\hat{Q} - \cdots \qquad (38b)$$

It was pointed out that the various $\hat{Q}$-boxes $\hat{Q}_1$, $\hat{Q}'$ and $\hat{Q}$ have <u>different</u> last-moment core insertions. As discussed above, last-moment core insertions result from attaching the interaction $H_1(V)$ at $t = 0$ to a valence line and the core wave function $U_Q(0,-\infty)|c\rangle$. Examples are the diagrams D1-D3 shown in fig. 17. These diagrams enter into $\hat{Q}_1$, $\hat{Q}'$ and $\hat{Q}$ according to the following rules [see the discussion of eqs. (37.5-7)]:

	D1	D2	D3
$\hat{Q}_1$	Yes	Yes	No
$\hat{Q}'$	No	Yes	No
$\hat{Q}$	Yes	Yes	Yes

(45)

Since $U_Q(0,-\infty)|c\rangle$ is evaluated <u>independently</u> of any valence particles (as is obvious from the decomposition theorem), we must be careful in the calculation of last-moment core-insertion diagrams. By ordinary diagram rules diagram D2 is given as

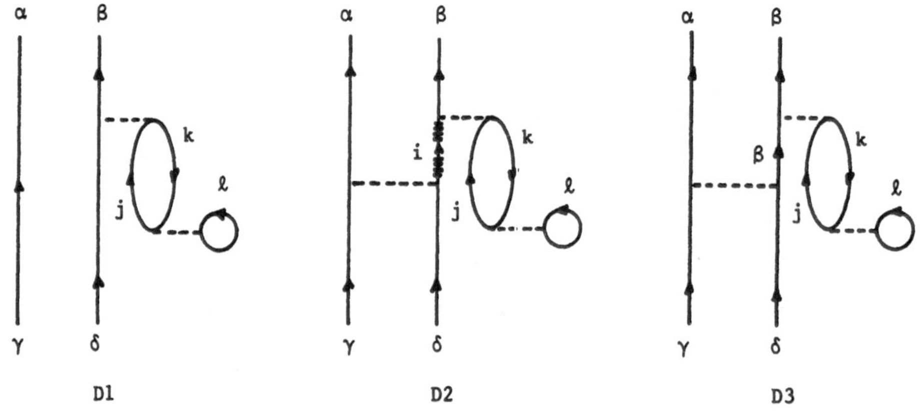

Fig. 17. Diagrams with last-moment core insertions.

$$D2 = \frac{1}{8} \frac{V_{\beta k,ij} V_{\alpha i,\gamma\delta} V_{j\ell,k\ell}}{[\varepsilon_\gamma+\varepsilon_\delta-(\varepsilon_\gamma+\varepsilon_\delta+\varepsilon_j-\varepsilon_k)][\varepsilon_\gamma+\varepsilon_\delta-(\varepsilon_\alpha+\varepsilon_i+\varepsilon_j-\varepsilon_k)]}$$

$$= \frac{1}{8} \frac{V_{\beta k,ij} V_{\alpha i,\gamma\delta} V_{j\ell,k\ell}}{(\varepsilon_k-\varepsilon_j)[\varepsilon_\gamma+\varepsilon_\delta-(\varepsilon_\alpha+\varepsilon_i+\varepsilon_j-\varepsilon_k)]} . \qquad (46)$$

Here, the core insertion is <u>off</u>-energy-shell, since the size of the diagram depends on where the core insertion is attached to the valence part of the diagram. This is not consistent with the fact that the core insertion is evaluated independently of the valence part and thus is <u>on</u>-energy-shell. Hence, the value of diagram D2 should be

$$D2 = \frac{1}{8} \frac{V_{\beta k,ij} V_{\alpha i,\gamma\delta} V_{j\ell,k\ell}}{(\varepsilon_k-\varepsilon_j)[\varepsilon_\gamma+\varepsilon_\delta-(\varepsilon_\alpha+\varepsilon_i)]} . \qquad (46.1a)$$

In order to obtain this, we must add two diagrams

D2 = 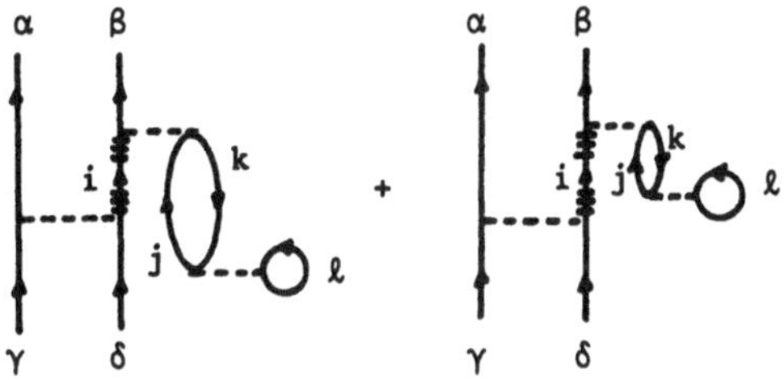

$$= \frac{1}{8} V_{\beta k,ij}\, V_{\alpha i,\gamma\delta}\, V_{j\ell,k\ell}$$

$$\times \left\{ \frac{1}{\varepsilon_\gamma+\varepsilon_\delta-(\varepsilon_\gamma+\varepsilon_\delta+\varepsilon_j-\varepsilon_k)}\; \frac{1}{\varepsilon_\gamma+\varepsilon_\delta-(\varepsilon_\alpha+\varepsilon_i+\varepsilon_j-\varepsilon_k)} \right.$$

$$\left. + \frac{1}{\varepsilon_\gamma+\varepsilon_\delta-(\varepsilon_\alpha+\varepsilon_i)}\; \frac{1}{\varepsilon_\gamma+\varepsilon_\delta-(\varepsilon_\alpha+\varepsilon_i+\varepsilon_j-\varepsilon_k)} \right\}$$

$$= \frac{1}{8} \frac{V_{\beta k,ij}\, V_{\alpha i,\gamma\delta}\, V_{j\ell,k\ell}}{(\varepsilon_k-\varepsilon_j)[\varepsilon_\gamma+\varepsilon_\delta-(\varepsilon_\alpha+\varepsilon_i)]} \;. \tag{46.1b}$$

This is just generalized time ordering of the interaction in the core insertion relative to the interaction in the valence part of the diagram.

In fact, all the $t = 0$ core insertions are <u>on</u>-energy-shell, because they arise from $U_Q(0,-\infty)\,|c\rangle$, which is evaluated independently from the valence particles. We denote these on-energy-shell core insertions by

$t = 0 \rightarrow$ 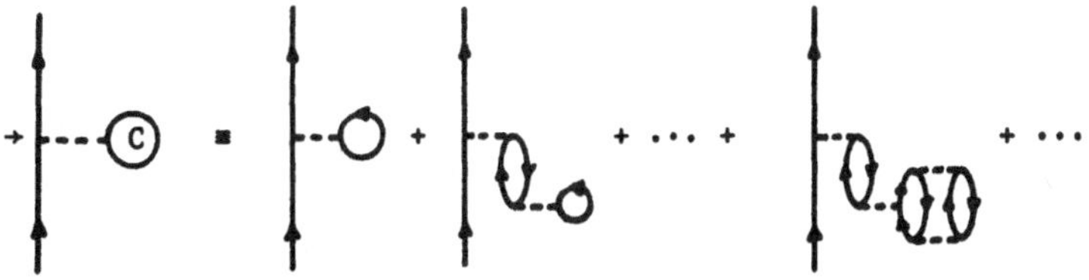 (47)

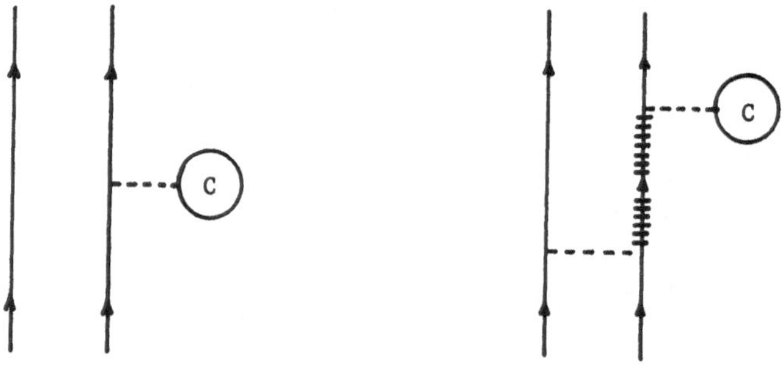

(a) (b)

Fig. 18. On-energy-shell core insertions.

Clearly, diagram D1 is of the general form shown in fig. 18a and diagram D2 of the general form shown in fig. 18b. Diagram D3 can also be made into the above form by way of g.t.o. As noted above, diagram D3 can only contribute to a $\hat{Q}$-box after folding (i.e. to the right of a fold). Consider for example the following diagrams before folding

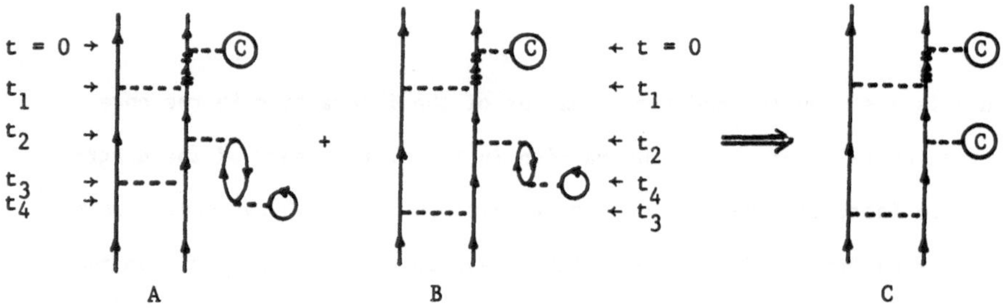

(48)

The lower part of diagram A (i.e. from t_2 to t_4) is just diagram D3 of fig. 17. Note that the top time of diagram D3 is before $t = 0$. According to the standard diagram rules the core insertion in diagram D3 is evaluated <u>off</u>-energy-shell. However, if the lower part of diagram B in eq. (48), which is an allowed diagram before folding, is also included in diagram D3, the core insertion will be <u>on</u>-energy-shell, as indicated by diagram C. But diagram C is a three-$\hat{Q}$-box term, which should be folded twice to give

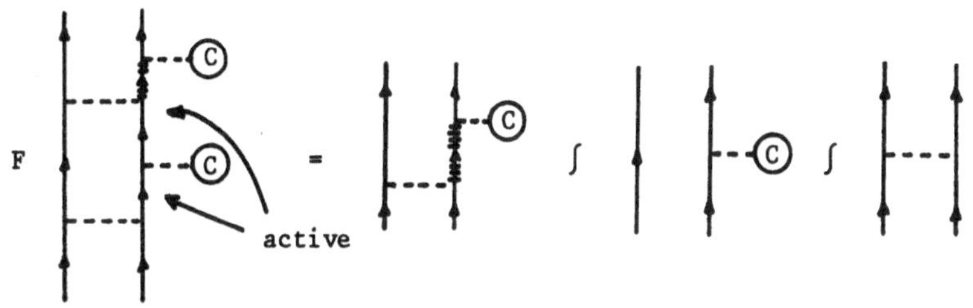

(48.1)

Thus, by putting the core insertions in diagrams like D3 on-energy-shell (by means of g.t.o.), these diagrams become reducible (because of the active intermediate state between the instantaneous core insertion and the last previous interaction in the valence part of the diagram) and must therefore be folded. Hence, in generalized time order, core-insertion diagrams of the type D3 do not contribute to $\hat{Q}$ (nor, of course, to $\hat{Q}_1$ and $\hat{Q}'$).

Then, we can write the expansion (38b) of the effective interaction in the form

$$v_{\text{eff}} = \hat{Q} - \hat{Q}'\int\hat{Q} + \hat{Q}'\int\hat{Q}\int\hat{Q} - \cdots , \tag{49}$$

where it is understood that the last-moment core insertions have been put on-energy-shell using g.t.o. Since diagrams of the type D3 have been eliminated, we have $\hat{Q}_1 = \hat{Q}$, and the effective interaction can be expressed in terms of only two different $\hat{Q}$-boxes $\hat{Q}$ and $\hat{Q}'$. These are related to each other by

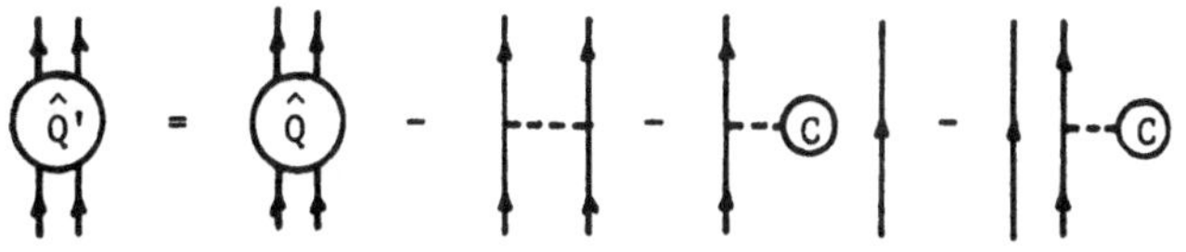

(50)

So far, only last-moment core insertions have been put on-energy-shell using g.t.o. It is convenient also to put core insertions at $t < 0$ on-energy-shell. An example is

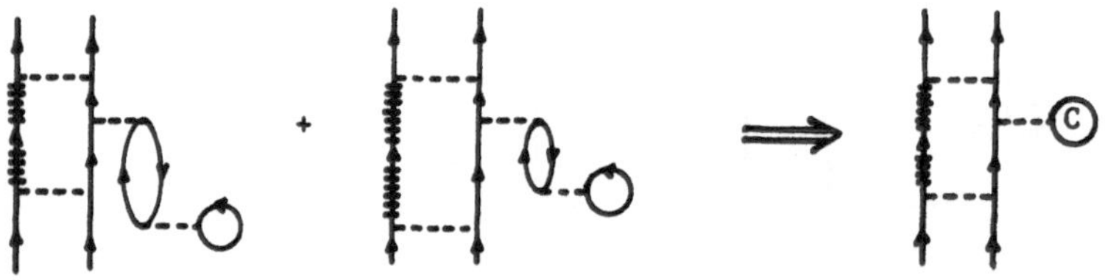

(51)

Note that the core insertions in eq. (51) are downward going. Upward going core insertions must, however, be treated off-energy-shell. An example is shown in fig. 19a. One might try to put the core insertion in fig. 19a on-energy-shell by adding the diagram shown in fig. 19b. This would give the energy denominator

$$\frac{1}{(\varepsilon_\gamma+\varepsilon_\delta)-(\varepsilon_i+\varepsilon_k)}\left\{\frac{1}{(\varepsilon_\gamma+\varepsilon_\delta)-(\varepsilon_i+\varepsilon_j+\varepsilon_\ell-\varepsilon_m)}\,\frac{1}{(\varepsilon_\gamma+\varepsilon_\delta)-(\varepsilon_i+\varepsilon_j)}\right.$$
$$\left.+\frac{1}{(\varepsilon_\gamma+\varepsilon_\delta)-(\varepsilon_i+\varepsilon_j+\varepsilon_\ell-\varepsilon_m)}\,\frac{1}{(\varepsilon_\gamma+\varepsilon_\delta)-(\varepsilon_\alpha+\varepsilon_\beta+\varepsilon_\ell-\varepsilon_m)}\right\}, \quad (52)$$

which would lead to the on-energy-shell expression

$$\frac{1}{(\varepsilon_\gamma+\varepsilon_\delta)-(\varepsilon_i+\varepsilon_k)}\,\frac{1}{(\varepsilon_\gamma+\varepsilon_\delta)-(\varepsilon_i+\varepsilon_j)}\,\frac{1}{-(\varepsilon_\ell-\varepsilon_m)} \quad (52.1)$$

only when $\varepsilon_\gamma+\varepsilon_\delta = \varepsilon_\alpha+\varepsilon_\beta$. However, when we use a model space consisting of several harmonic oscillator shells, we may have diagrams with $\varepsilon_\gamma+\varepsilon_\delta \neq \varepsilon_\alpha+\varepsilon_\beta$. Thus, in general, upward going insertions are off-energy-shell.†

---

† Note that in the generalized folded diagram theory of Kuo and Krenciglowa[65], upward going core insertions can be made on-energy-shell.

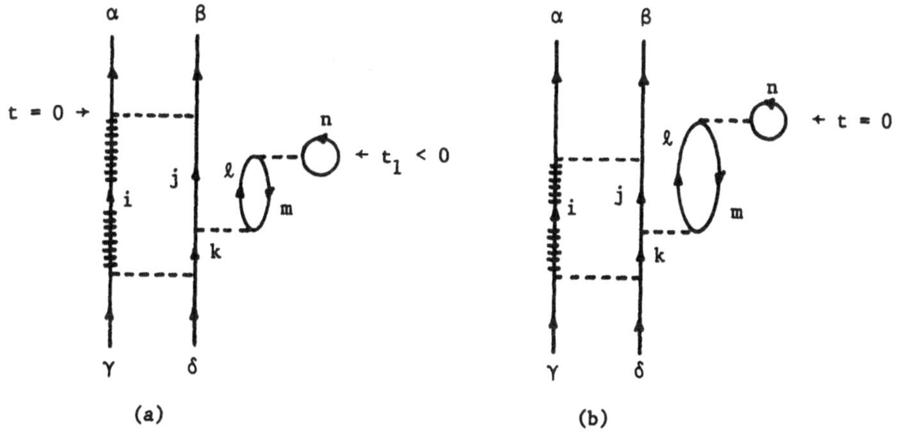

Fig. 19. Diagrams with upward going core insertions. As discussed in the text, upward going core insertions will be treated as <u>off</u>-energy-shell core insertions.

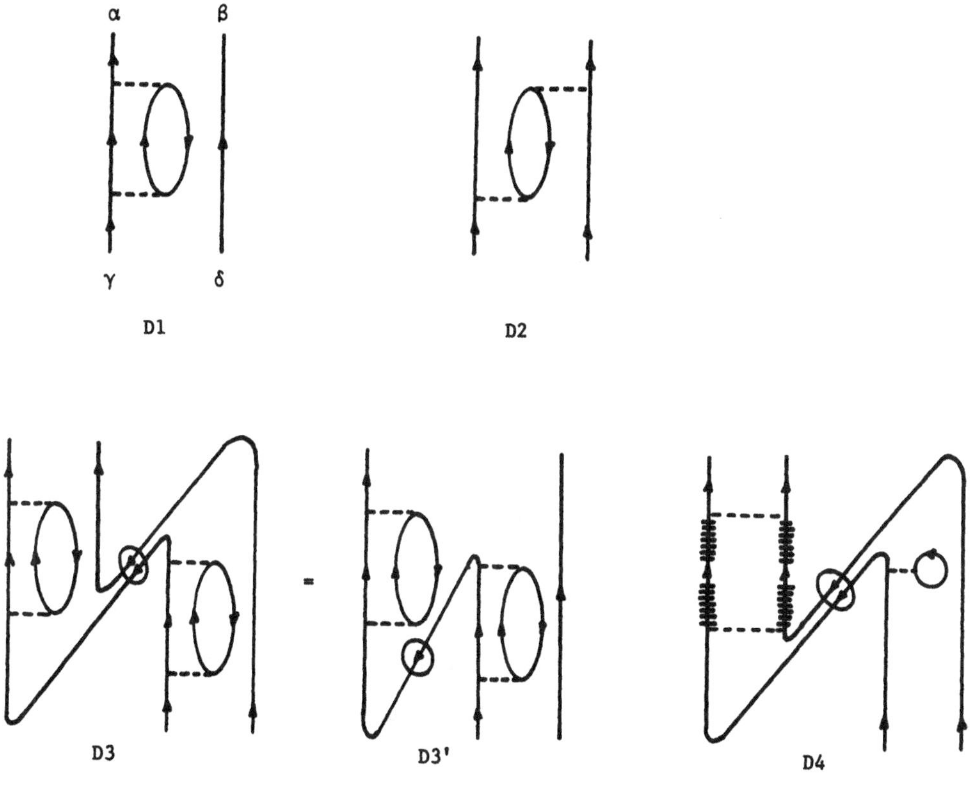

Fig. 20. Diagrams belonging to $\hat{Q}$ (D1-D2) and $\hat{Q}'\int\hat{Q}$ (D3-D4).

<u>5.3. Many-body effective interactions.</u> Although the interaction V of eq. (6.3) used in the original Hamiltonian is purely two-body, the P-space effective interaction v_{eff} will contain many-body forces of rank n, with $n \leq N_V$, where N_V is the number of valence nucleons. Consider first the case of two valence particles which has already been discussed quite extensively. In fig. 20 we show various diagrams contributing to v_{eff} of eq. (49). Here, diagrams D1 and D2 belong to $\hat{Q}$, whereas diagrams D3 and D4 belong to $\hat{Q}'\int\hat{Q}$. Clearly, diagram D1 has one-body character, since only one valence line has interactions attached to it. Similarly, the folded diagram D3 is one-body. On the other hand, diagrams D2 and D4 are two-body, since both valence lines are linked together by interactions. (Note that, since we do not have disconnected diagrams, all valence lines which have interactions attached to them must form <u>one</u> totally connected piece.) If there were disconnected diagrams in v_{eff}, we would not be able to decompose v_{eff} in the way discussed below. To summarize, in the case of two valence nucleons, we can decompose v_{eff} as follows

$$v_{\text{eff}} = v_{\text{eff}}(1) + v_{\text{eff}}(2) , \tag{53}$$

where $v_{\text{eff}}(1)$ is the sum of all one-body diagrams (such as D1 and D3 in fig. 20) and $v_{\text{eff}}(2)$ is the sum of all two-body diagrams (such as D2 and D4 in fig. 20).

Similarly, in the case of N_V valence nucleons we have

$$v_{\text{eff}} = v_{\text{eff}}(1) + v_{\text{eff}}(2) + \cdots + v_{\text{eff}}(n) + \cdots + v_{\text{eff}}(N_V) , \tag{54}$$

where $v_{\text{eff}}(n)$ represents the n-body effective interaction. In fig. 21 we show a three-body and a four-body diagram contributing to the effective interaction in a system of four valence nucleons.

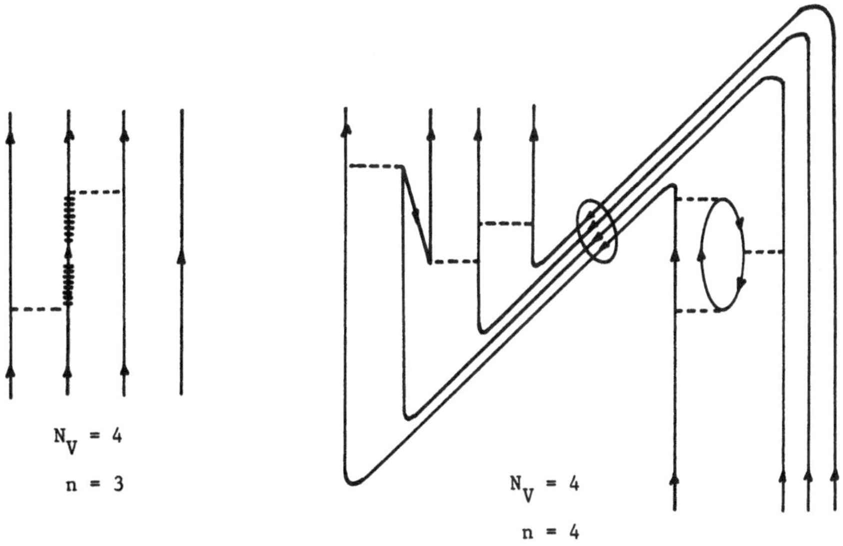

Fig. 21. Three- and four-body diagrams with four valence nucleons.

The decomposition of v_{eff} into parts $v_{eff}(n)$ of different particle rank can be very useful. Suppose we are calculating ^{18}O, which has two valence neutrons. Then, v_{eff} is composed of $v_{eff}(1)$ and $v_{eff}(2)$. We can show that $v_{eff}(1)$ may be obtained from the observed properties of ^{17}O and ^{16}O. Thus, starting from known properties of ^{16}O and ^{17}O, we can calculate ^{18}O merely by evaluating the two-body effective interaction $v_{eff}(2)$.

In order to see this, let us first consider a P-space calculation of ^{17}O. Take a one-dimensional P-space

$$P = |\Phi\rangle\langle\Phi| , \tag{55}$$

with

$$|\Phi\rangle = a^{\dagger}_{j_1} |c\rangle , \qquad j_1 \equiv (n_1 \ell_1 j_1) = 0d_{5/2} . \tag{55.1}$$

Then, we have a one-dimensional P-space secular equation and readily obtain

$E(^{17}O;\ g.s.) - E(^{16}O;\ g.s.)$

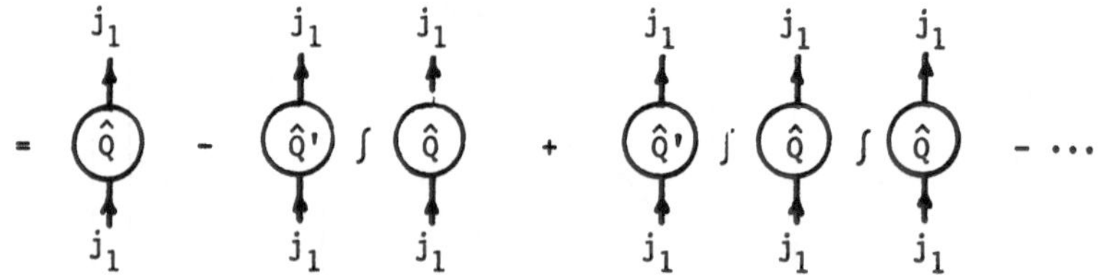

(56)

Typical diagrams in this expansion are

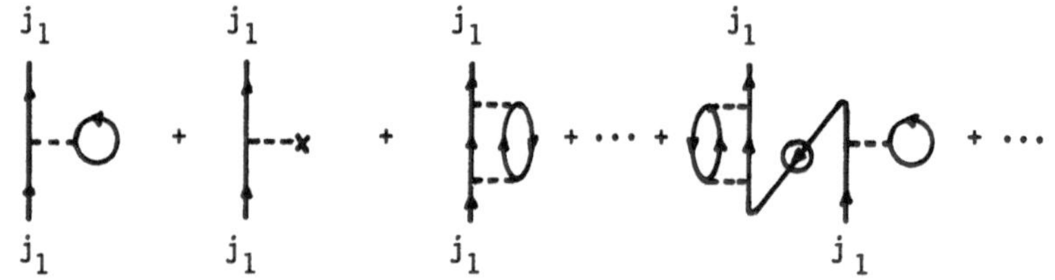

(56.1)

Clearly, all the above one-body diagrams continue to have the same value if we add an idle line to them, e.g.

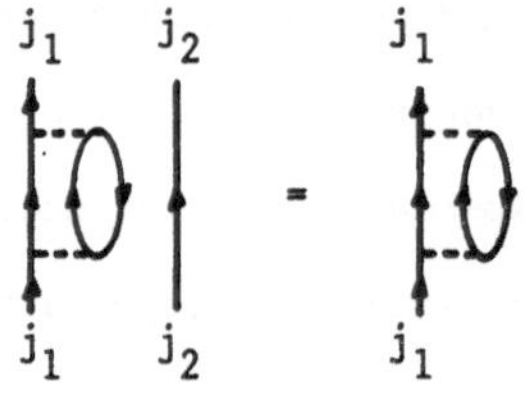

(57)

Thus, for the $\{0d_{5/2}, 1s_{1/2}, 0d_{3/2}\}$ model space, all the one-body diagrams of ^{18}O are identical to the one-body diagrams of ^{17}O, the sum of which may be extracted from experimental binding energy differences as shown by eq. (56). (Clearly, eq. (56) applies to $j_2 = 1s_{1/2}$ and $j_3 = 0d_{3/2}$ as well.) Then, the effective Hamiltonian to be used in the secular equation

$$H_{eff}P\Psi_\lambda = (E_\lambda - E_C)P\Psi_\lambda \tag{58}$$

for ^{18}O in the $\{0d_{5/2}, 1s_{1/2}, 0d_{3/2}\}$ model space can be written as

$$H_{eff} = \tilde{H}_o(V) + v_{eff}(2) , \tag{58.1}$$

where

$$\tilde{H}_o(V) = H_o(V) + v_{eff}(1) = \sum_j \tilde{\varepsilon}_j a_j^\dagger a_j . \tag{58.2}$$

Here,

$$\tilde{\varepsilon}_j = E(^{17}O;\ j) - E(^{16}O;\ j) \tag{58.3}$$

is the difference in binding energies of the lowest state in ^{17}O with angular momentum j (and non-zero overlap with the single-particle state j) and the ground state in ^{16}O. The effective two-body interaction in eq. (58.1) is written as

$$v_{eff}(2) = \frac{1}{4} \sum_{\alpha\beta\gamma\delta} [v_{eff}(2)]_{\alpha\beta,\gamma\delta}\ a_\alpha^\dagger a_\beta^\dagger a_\delta a_\gamma , \tag{58.4}$$

where

$[v_{eff}(2)]_{\alpha\beta,\gamma\delta}$ =

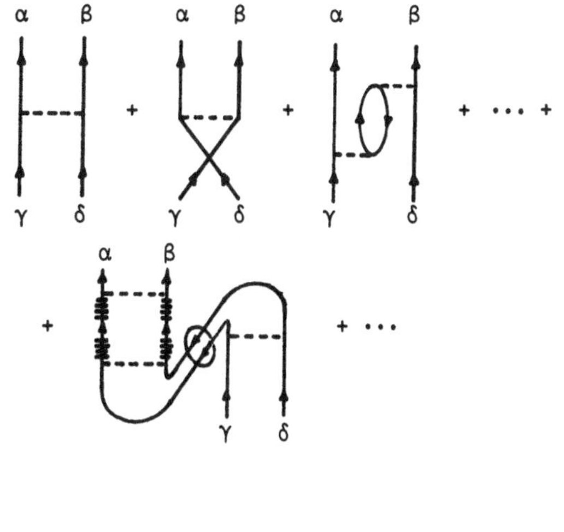

(58.5)

is the sum of all the irreducible non-folded and folded two-body diagrams of v_{eff}.

Thus, we have obtained a procedure for evaluating the spectrum of ^{18}O which is very closely analogous to the highly successful empirical shell-model approach. The above procedure can be summarized as follows: Once the ^{17}O problem has been successfully solved, we can use the calculated one-body energies (or if these are not well determined, the corresponding experimental energies) in the ^{18}O problem. Then, only the two-body effective interaction remains to be calculated. (Recall that in the empirical shell-model approach experimental single-particle energies are used, and the two-body effective interaction is parametrized.)

The above folded diagram approach allows us to consider a complex nucleus as composed of neighbouring nuclei of lower mass number as building blocks. For example, the building blocks of ^{18}O are ^{16}O and ^{17}O, and the remaining ingredient in the calculation of the ^{18}O spectrum is the two-body effective interaction $v_{eff}(2)$. Similarly, ^{20}Ne may be thought of as being built up from ^{16}O, ^{18}O, ^{18}F and ^{18}Ne and the three- and four-body effective interactions $v_{eff}(3)$ and $v_{eff}(4)$. The latter case may need some explanation. The contributions from $H_o(V) + v_{eff}(1) + v_{eff}(2)$ are shown schematically in fig. 22 and may be extracted from the binding energy differences between the appropriate mass 18 nuclei and ^{16}O. However, we really need off-diagonal matrix elements corresponding to the diagrams shown in fig. 22. In order to extract matrix elements from experimental binding energy differences $E_\lambda - E_C$, it is necessary to know the corresponding wave-function projections $P\Psi_\lambda$. This is because the effective Hamiltonian for, say, ^{18}O can be written as

$$H_{eff}(^{18}O) = \sum_\lambda |\overline{P\Psi_\lambda}\rangle(E_\lambda - E_C)\langle P\Psi_\lambda| , \tag{59}$$

where $\overline{P\Psi_\lambda}$ is biorthogonal to $P\Psi_\lambda$. Thus, knowing all the projections

$P\Psi_\lambda$, we can construct all the P-space matrix elements of $H_{eff}(^{18}O)$. Usually we cannot determine $P\Psi_\lambda$ from experimental data except in the most simple case where the P-space is one-dimensional (see e.g. the above determination of one-body matrix elements). Instead, we may determine $P\Psi_\lambda$ approximately, for example by calculating it using an approximate v_{eff} obtained from low-order perturbation theory.

Fig. 22. Contributions to ^{20}Ne from $H_o(V) + v_{eff}(1) + v_{eff}(2)$ which in principle can be extracted from the binding energies of ^{18}Ne, ^{18}F and ^{18}O, respectively.

5.4. Folded diagrams and energy derivatives of $\hat{Q}$-boxes. Here, we discuss a convenient method for calculating folded diagrams. A preliminary discussion of this method was given in sect. 2.

We begin by simple examples. First, consider the diagram

D1 =

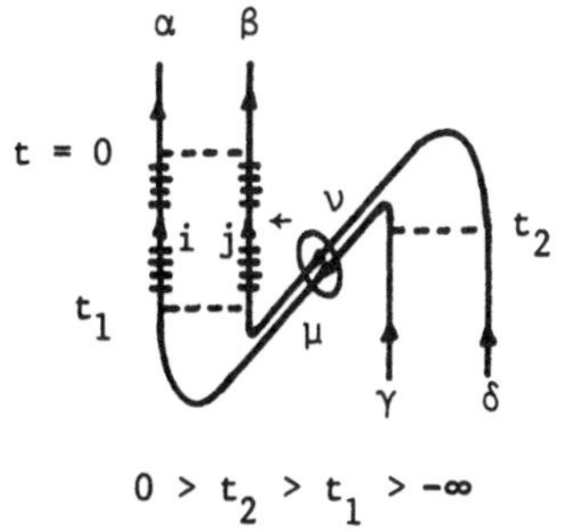

$$= \frac{1}{8} \frac{V_{\alpha\beta,ij} V_{ij,\mu\nu} V_{\mu\nu,\gamma\delta}}{(\varepsilon_\gamma+\varepsilon_\delta-\varepsilon_i-\varepsilon_j)(\varepsilon_\mu+\varepsilon_\nu-\varepsilon_i-\varepsilon_j)}$$

$$= \frac{1}{8} \frac{V_{\alpha\beta,ij} V_{ij,\mu\nu} V_{\mu\nu,\gamma\delta}}{(\varepsilon_\gamma+\varepsilon_\delta-\varepsilon_i-\varepsilon_j)^2} , \tag{60a}$$

where we have assumed a degenerate model space, i.e. $\varepsilon_\mu+\varepsilon_\nu = \varepsilon_\gamma+\varepsilon_\delta$. The arrow indicates that the interaction at the time t_2 to the right of the fold is between the interactions at the times $t = 0$ and t_1 to the left of the fold. For a degenerate P-space the above diagram can be written as

$$D1 = - [\frac{d}{d\omega} \left\{ \frac{1}{4} \frac{V_{\alpha\beta,ij} V_{ij,\mu\nu}}{\omega-\varepsilon_i-\varepsilon_j} \right\}_{\omega=\varepsilon_\gamma+\varepsilon_\delta}] \times \frac{1}{2} V_{\mu\nu,\gamma\delta} . \tag{60b}$$

As another example, consider

D2 = 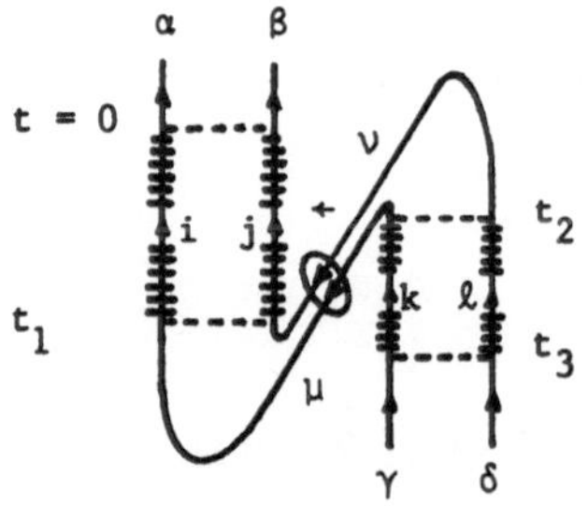

$0 > t_2 > t_1 > t_3 > -\infty$

$0 > t_2 > t_3 > t_1 > -\infty$

$$= \frac{1}{16} V_{\alpha\beta,ij} V_{ij,\mu\nu} V_{\mu\nu,k\ell} V_{k\ell,\gamma\delta}$$

$$\times \left\{ \frac{1}{(\varepsilon_\gamma+\varepsilon_\delta-\varepsilon_i-\varepsilon_j)(\varepsilon_\gamma+\varepsilon_\delta+\varepsilon_\mu+\varepsilon_\nu-\varepsilon_i-\varepsilon_j-\varepsilon_k-\varepsilon_\ell)(\varepsilon_\gamma+\varepsilon_\delta-\varepsilon_k-\varepsilon_\ell)} \right.$$

$$\left. + \frac{1}{(\varepsilon_\gamma+\varepsilon_\delta-\varepsilon_i-\varepsilon_j)(\varepsilon_\gamma+\varepsilon_\delta+\varepsilon_\mu+\varepsilon_\nu-\varepsilon_i-\varepsilon_j-\varepsilon_k-\varepsilon_\ell)(\varepsilon_\mu+\varepsilon_\nu-\varepsilon_i-\varepsilon_j)} \right\}$$

$$= \frac{1}{16} \frac{V_{\alpha\beta,ij} V_{ij,\mu\nu} V_{\mu\nu,k\ell} V_{k\ell,\gamma\delta}}{(\varepsilon_\gamma+\varepsilon_\delta-\varepsilon_i-\varepsilon_j)(\varepsilon_\gamma+\varepsilon_\delta-\varepsilon_k-\varepsilon_\ell)(\varepsilon_\mu+\varepsilon_\nu-\varepsilon_i-\varepsilon_j)}$$

$$= \frac{1}{16} \frac{V_{\alpha\beta,ij} V_{ij,\mu\nu} V_{\mu\nu,k\ell} V_{k\ell,\gamma\delta}}{(\varepsilon_\gamma+\varepsilon_\delta-\varepsilon_i-\varepsilon_j)^2(\varepsilon_\gamma+\varepsilon_\delta-\varepsilon_k-\varepsilon_\ell)} \quad . \tag{61a}$$

Here, the arrow denotes a generalized time ordering (g.t.o.) where t_2 is fixed relative to $t = 0$ and t_1, but t_3 is free to move up and down. Thus, diagram D2 actually consists of two diagrams corresponding to whether t_3 is before or after t_1. In the last step of eq. (61a) we have again assumed degenerate P-space energies. With this assumption we can write diagram D2 as

$$D2 = -\left[\frac{d}{d\omega} \left\{ \frac{1}{4} \frac{V_{\alpha\beta,ij} V_{ij,\mu\nu}}{\omega-\varepsilon_i-\varepsilon_j} \right\}_{\omega=\varepsilon_\gamma+\varepsilon_\delta} \right] \times \left\{ \frac{1}{4} \frac{V_{\mu\nu,k\ell} V_{k\ell,\gamma\delta}}{\varepsilon_\gamma+\varepsilon_\delta-\varepsilon_k-\varepsilon_\ell} \right\} . \tag{61b}$$

This result can be interpreted in the following way. By g.t.o. we have released the time constraints on t_3 (except that it must be before t_2); thus the t_2 to t_3 part of the diagram behaves like an instantaneous interaction. Hence, diagram D2 can be obtained from diagram D1 by replacing $V_{\mu\nu,\gamma\delta}$ by $\frac{1}{4} V_{\mu\nu,k\ell}V_{k\ell,\gamma\delta}/(\varepsilon_\gamma+\varepsilon_\delta-\varepsilon_k-\varepsilon_\ell)$.

The above can be generalized. Write the $\hat{Q}$-box as

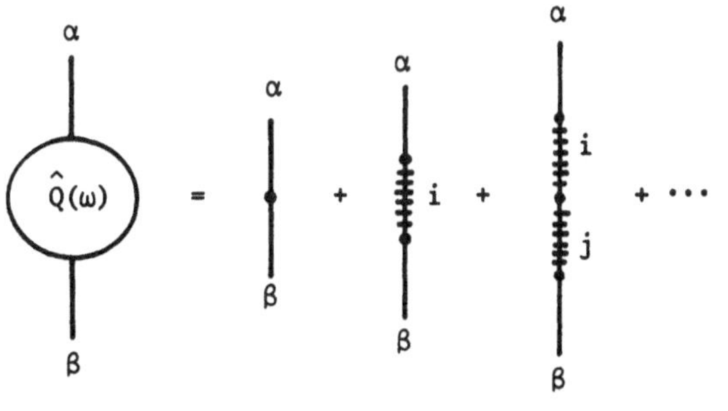

(62a)

$$= V + \sum_i \frac{V_{\alpha i} V_{i\beta}}{\omega - W_i} + \sum_{ij} \frac{V_{\alpha i} V_{ij} V_{j\beta}}{(\omega - W_i)(\omega - W_j)} + \cdots \tag{62b}$$

$$= \left[V + V \frac{Q_{LV}}{\omega - H_o} V + V \frac{Q_{LV}}{\omega - H_o} V \frac{Q_{LV}}{\omega - H_o} V + \cdots \right]_{\alpha\beta} \tag{62c}$$

In eq. (62a) simple lines (|) denote active states, railed lines (⧘) denote passive states (i.e. containing at least one passive line) and dots (·) denote H_1 vertices and downward going on-energy-shell core insertions [for simplicity we take $H_1 = V$ in eqs. (62b-c)].† It is understood that all the terms in eq. (62) are valence-linked, i.e. all the vertices are attached to at least one active particle line. Examples of diagrams contributing to eq. (62a) are shown in fig. 23. Note that the vertices can be attached directly (e.g. vertices a, c and d of diagram C in fig. 23) or indirectly (e.g. vertex b of diagram C in fig. 23) to active lines. In eq. (62b) W_i is the eigenvalue of H_o (i.e. the sum of unperturbed single-particle energies) in the state i. In eq. (62c) Q_{LV} stands for summation over passive intermediate states leading to valence-linked terms. In the present formulation of the $\hat{Q}$-box, $\hat{Q}'$ is equal to $\hat{Q}$ minus the first term.

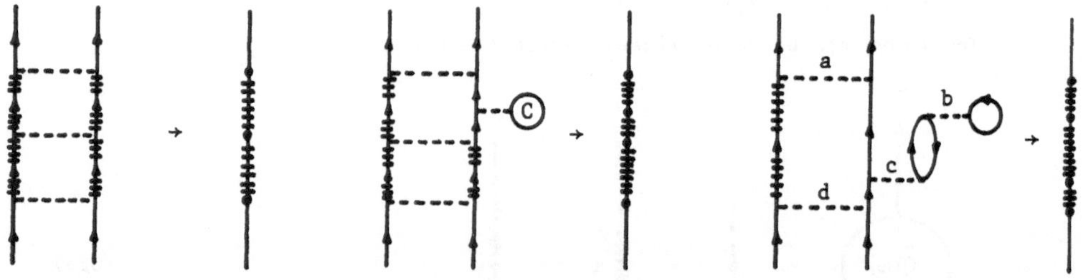

A B C

Fig. 23. Examples of diagrams contributing to $\hat{Q}$ of eq. (62).

---

† This is a slight simplification of the notation (15) used in previous sections.

It is now easily verified from the general definition of $\hat{Q}$ in eq. (62b) that

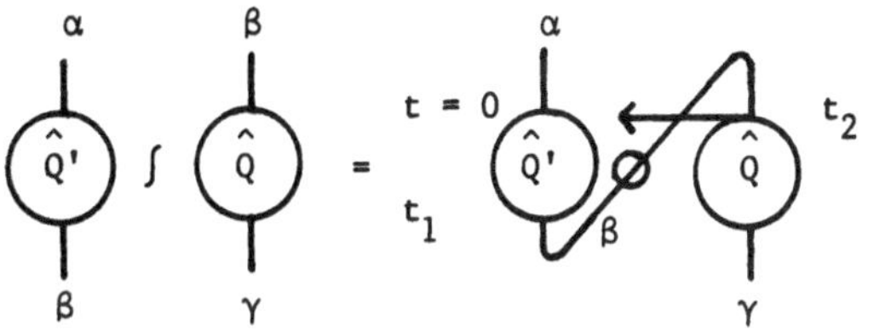

$$= - \langle\alpha|[\frac{d\hat{Q}'(\omega)}{d\omega}]_{\omega=W_\gamma}|\beta\rangle\langle\beta|\hat{Q}(W_\gamma)|\gamma\rangle \,. \qquad (63)$$

Here, t_2 is the top time of $\hat{Q}$ which must be between the top and bottom times $t = 0$ and t_1 of $\hat{Q}'$. Furthermore, generalized time ordering is assumed, implying that all diagrams with $0 > t_2 > t_1$ are included. The result of eq. (63) can also be obtained from the defining equation of folded $\hat{Q}$-box diagrams [cf. eq. (10)]:

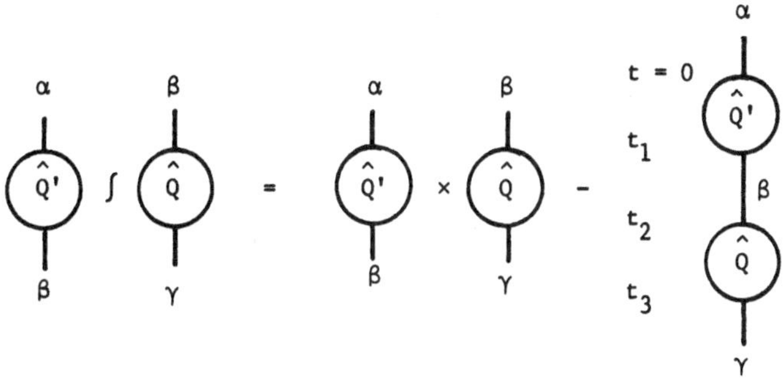

$$= \frac{\langle\alpha|\hat{Q}'(W_\beta)|\beta\rangle\langle\beta|\hat{Q}(W_\gamma)|\gamma\rangle}{W_\gamma - W_\beta} - \frac{\langle\alpha|\hat{Q}'(W_\gamma)|\beta\rangle\langle\beta|\hat{Q}(W_\gamma)|\gamma\rangle}{W_\gamma - W_\beta}$$

$$= - \frac{\langle\alpha|\hat{Q}'(W_\gamma)|\beta\rangle - \langle\alpha|\hat{Q}'(W_\beta)|\beta\rangle}{W_\gamma - W_\beta} \langle\beta|\hat{Q}(W_\gamma)|\gamma\rangle \qquad (63.1a)$$

$$= - \langle\alpha|[\frac{d\hat{Q}'(\omega)}{d\omega}]_{\omega=W_\gamma}|\beta\rangle\langle\beta|\hat{Q}(W_\gamma)|\gamma\rangle \,. \qquad (63.1b)$$

Here, the last step was made assuming a degenerate model space, i.e. $W_\gamma - W_\beta \to 0$.

We can in fact express all folded diagrams as derivatives of $\hat{Q}$-boxes. Consider for example the twice-folded diagram

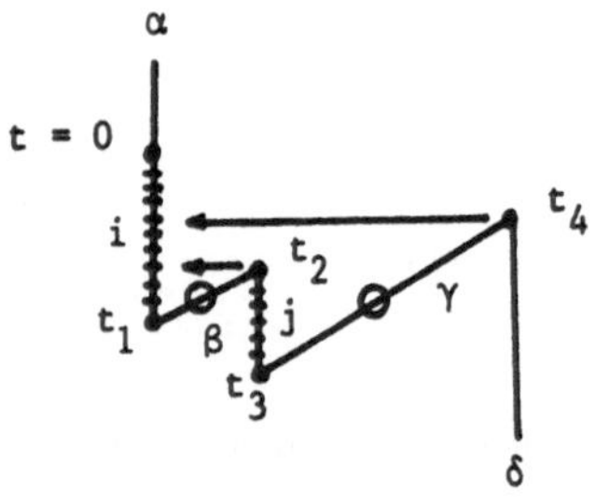

$0 > t_4 > t_2 > t_1$

$$= \frac{V_{\alpha i} V_{i\beta} V_{\beta j} V_{j\gamma} V_{\gamma\delta}}{(W_\delta - W_i)(W_\gamma - W_i)(W_\beta - W_i)(W_\gamma - W_j)} = \frac{V_{\alpha i} V_{i\beta} V_{\beta j} V_{j\gamma} V_{\gamma\delta}}{(W_\delta - W_i)^3 (W_\delta - W_j)}$$

$$= \left[\frac{1}{2!} \frac{d^2}{d\omega^2} \left\{ \frac{V_{\alpha i} V_{i\beta}}{\omega - W_i} \right\}_{\omega = W_\delta} \right] \times \left\{ \frac{V_{\beta j} V_{j\gamma}}{W_\delta - W_j} \right\} \times V_{\gamma\delta} \, . \tag{64}$$

Again, we have assumed g.t.o. (i.e. t_3 can be before and after t_1) and degenerate P-space (i.e. $W_\delta = W_\gamma = W_\beta$). Similarly, we obtain for a different type of twice-folded diagram

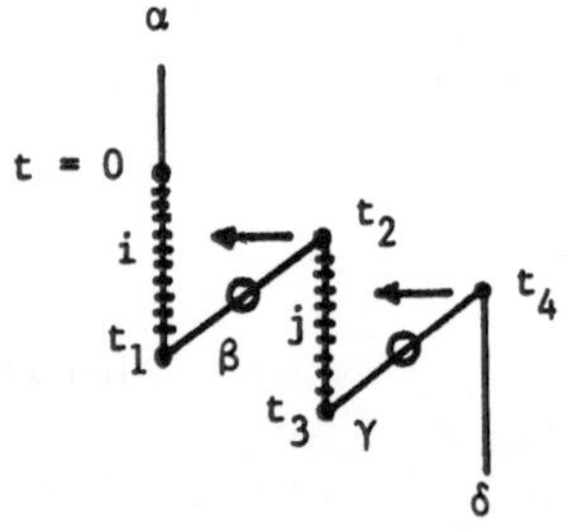

$0 > t_2 > t_1$

$t_2 > t_4 > t_3$

$$= \frac{V_{\alpha i} V_{i\beta} V_{\beta j} V_{j\gamma} V_{\gamma\delta}}{(W_\delta - W_i)(W_\beta - W_i)(W_\delta - W_j)(W_\gamma - W_j)} = \frac{V_{\alpha i} V_{i\beta} V_{\beta j} V_{j\gamma} V_{\gamma\delta}}{(W_\delta - W_i)^2 (W_\delta - W_j)^2}$$

$$= [\frac{d}{d\omega} \left\{ \frac{V_{\alpha i} \; V_{i\beta}}{\omega - W_i} \right\}_{\omega = W_\delta}] \times [\frac{d}{d\omega} \left\{ \frac{V_{\beta j} \; V_{j\gamma}}{\omega - W_j} \right\}_{\omega = W_\delta}] \times V_{\gamma\delta} \; . \qquad (65)$$

As in eq. (64) we have used g.t.o. and degenerate P-space.

The diagrams shown in eqs. (64) and (65) can be generalized to folded $\hat{Q}$-box diagrams of the types

D1 =

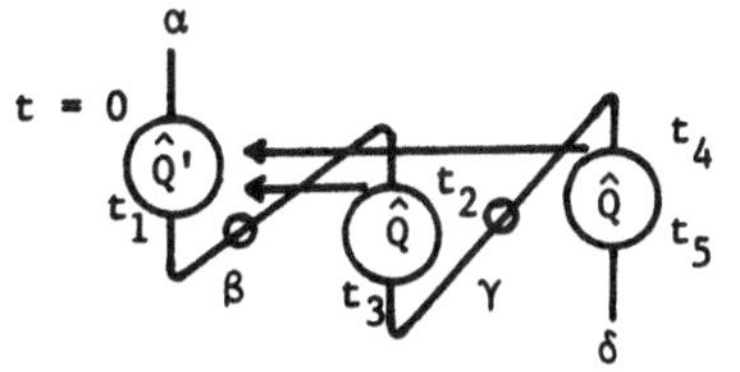

$$= \langle\alpha| \frac{1}{2!} [\frac{d^2\hat{Q}'(\omega)}{d\omega^2}]_{\omega=W_\delta} |\beta\rangle\langle\beta|\hat{Q}(W_\delta)|\gamma\rangle\langle\gamma|\hat{Q}(W_\delta)|\delta\rangle \qquad (66)$$

and

D2 =

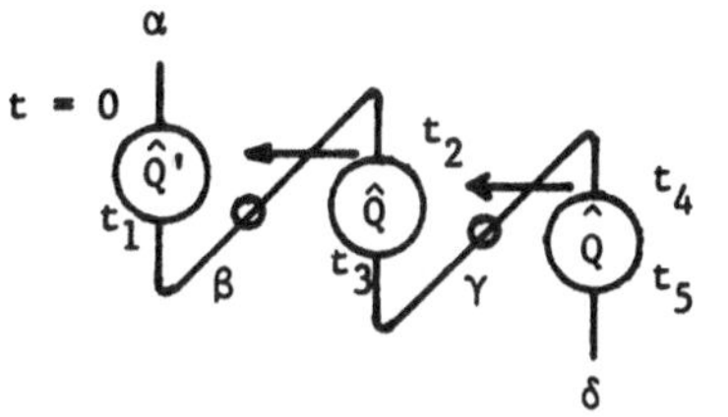

$$= \langle\alpha| [\frac{d\hat{Q}'(\omega)}{d\omega}]_{\omega=W_\delta} |\beta\rangle\langle\beta| [\frac{d\hat{Q}(\omega)}{d\omega}]_{\omega=W_\delta} |\gamma\rangle\langle\gamma|\hat{Q}(W_\delta)|\delta\rangle \, , \qquad (67)$$

respectively. The time constraints on these diagrams are

$$0 > t_4 > t_2 > t_1, \quad t_2 > t_3, \quad t_4 > t_5 \qquad \text{for D1,}$$

$$0 > t_2 > t_1, \quad t_2 > t_4 > t_3, \quad t_4 > t_5 \qquad \text{for D2.} \qquad (68)$$

Thus, the total contribution from D1 and D2 will have the following time constraints

$$0 > t_2 > t_1, \quad 0 > t_4 > t_3, \quad t_2 > t_3, \quad t_4 > t_5 . \tag{69}$$

These are the time constraints satisfied by the generalized twice-folded contribution

D3 =

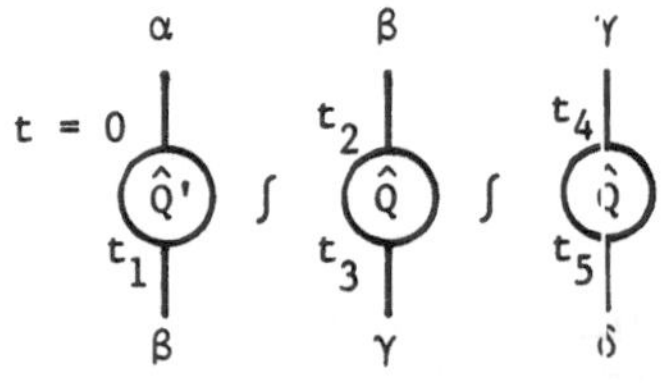

= D1 + D2

$$= \frac{1}{2!} \frac{d^2\hat{Q}'}{d\omega^2} P\hat{Q}P\hat{Q} + \frac{d\hat{Q}'}{d\omega} P \frac{d\hat{Q}}{d\omega} P\hat{Q} . \tag{70}$$

Here, we have substituted the results of eqs. (66) and (67) for D1 and D2 on operator form.

The result of eq. (70) can also be derived from the defining equation for D3, as was done for the generalized once-folded $\hat{Q}$-box diagram in eq. (63.1). We derive D3 diagrammatically by factorization of a three-$\hat{Q}$-box diagram

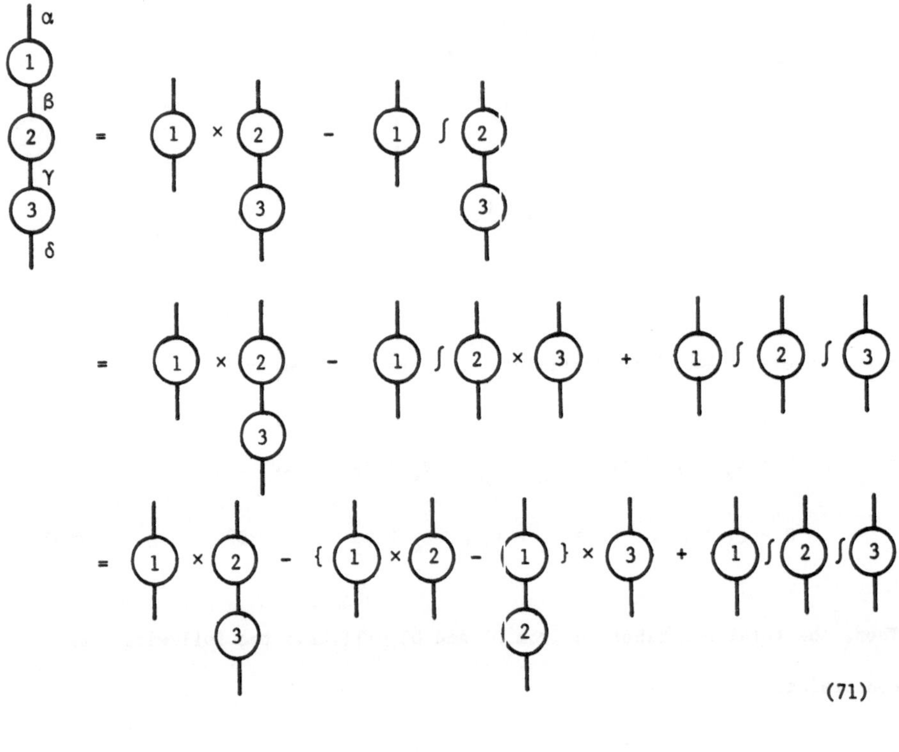

(71)

Here, the various terms can be written as

$$= \frac{\langle\alpha|\hat{Q}'(W_\delta)|\beta\rangle\langle\beta|\hat{Q}(W_\delta)|\gamma\rangle\langle\gamma|\hat{Q}(W_\delta)|\delta\rangle}{(W_\delta-W_\beta)(W_\delta-W_\gamma)}\,, \tag{71.1}$$

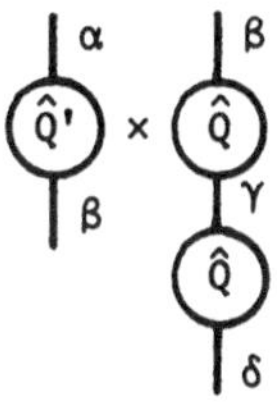

$$= \frac{\langle\alpha|\hat{Q}'(W_\beta)|\beta\rangle\langle\beta|\hat{Q}(W_\delta)|\gamma\rangle\langle\gamma|\hat{Q}(W_\delta)|\delta\rangle}{(W_\delta-W_\beta)(W_\delta-W_\gamma)}\,, \tag{71.2}$$

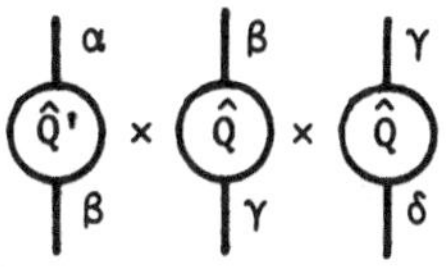

$$= \frac{\langle\alpha|\hat{Q}'(W_\beta)|\beta\rangle\langle\beta|\hat{Q}(W_\gamma)|\gamma\rangle\langle\gamma|\hat{Q}(W_\delta)|\delta\rangle}{(W_\gamma-W_\beta)(W_\delta-W_\gamma)}\,, \tag{71.3}$$

and

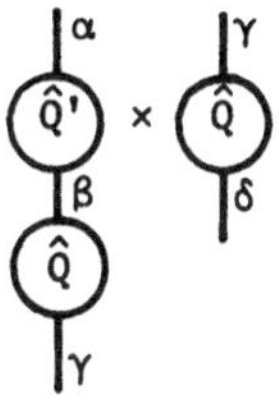

$$= \frac{\langle\alpha|\hat{Q}'(W_\gamma)|\beta\rangle\langle\beta|\hat{Q}(W_\gamma)|\gamma\rangle\langle\gamma|\hat{Q}(W_\delta)|\delta\rangle}{(W_\gamma-W_\beta)(W_\delta-W_\gamma)}\,. \tag{71.4}$$

Here, summation over the P-space intermediate states β and γ is implied. From eqs. (71.1-4) we readily obtain

D3 =

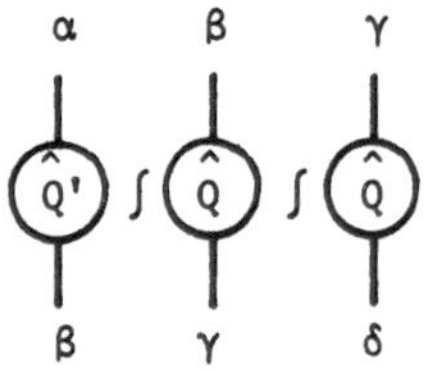

$$= \frac{\langle\alpha| \dfrac{\hat{Q}'(W_\delta)-\hat{Q}'(W_\beta)}{W_\delta-W_\beta} |\beta\rangle\langle\beta|\hat{Q}(W_\delta)|\gamma\rangle - \langle\alpha| \dfrac{\hat{Q}'(W_\gamma)-\hat{Q}'(W_\beta)}{W_\gamma-W_\beta} |\beta\rangle\langle\beta|\hat{Q}(W_\gamma)|\gamma\rangle}{W_\delta-W_\gamma}$$

$$\times \langle\gamma|\hat{Q}(W_\delta)|\delta\rangle \tag{71.5a}$$

$$= \left\{ \left[\frac{f(W_\delta)-f(W_\beta)}{W_\delta-W_\beta} \right] g(W_\delta) - \left[\frac{f(W_\gamma)-f(W_\beta)}{W_\gamma-W_\beta} \right] g(W_\gamma) \right\} \frac{g(W_\delta)}{W_\delta-W_\gamma} . \tag{71.5b}$$

Since we are assuming degenerate P-space, we take the limit of eq. (71.5b) as $W_\beta \to W_\gamma \to W_\delta$. By making Taylor expansions of f and g and keeping only low-order terms, we obtain for eq. (71.5b)

$$D3 = \{[f'(W_\beta) + \frac{1}{2!} f''(W_\beta)(W_\delta-W_\beta)]g(W_\delta) - [f'(W_\beta) + \frac{1}{2!} f''(W_\beta)(W_\gamma-W_\beta)]g(W_\gamma)\}$$

$$\times \frac{g(W_\delta)}{W_\delta-W_\gamma} \tag{71.5c}$$

$$= \left\{ f'(W_\beta)\left[\frac{g(W_\delta)-g(W_\gamma)}{W_\delta-W_\gamma} \right] + \frac{1}{2!} f''(W_\beta)\left[\frac{(W_\delta-W_\beta)g(W_\delta)-(W_\gamma-W_\beta)g(W_\gamma)}{W_\delta-W_\gamma} \right]\right\} g(W_\delta)$$

$$= f'(W_\beta)g'(W_\gamma)g(W_\delta) + \frac{1}{2!} f''(W_\beta)g(W_\gamma)g(W_\delta) , \qquad W_\beta = W_\gamma = W_\delta . \tag{71.5d}$$

Clearly, this result agrees with eq. (70).

It is interesting as well as important to note the following point. The r.h.s. of eqs. (71.1-4) are each divergent when we have a degenerate model space, i.e. $W_\beta = W_\gamma = W_\delta$. But the folded diagram D3, which is given by the combination (71.1) - (71.2) + (71.3) - (71.4), is not divergent. In this combination the singularities due to the vanishing of the energy denominators cancel exactly among themselves.

The above results can be extended to the general case of n folds. In an n-folded $\hat{Q}$-box diagram there are of course n+1 $\hat{Q}$-boxes. If all the n $\hat{Q}$-boxes to the right of $\hat{Q}'$ are folded into $\hat{Q}'$ (as indicated by <u>all</u> the arrows pointing towards $\hat{Q}'$), we obtain

$$\frac{1}{n!}\frac{d^n\hat{Q}'}{d\omega^n} P\hat{Q}P\hat{Q}P \cdots \hat{Q} . \tag{72}$$

Quite generally, the number of arrows pointing towards a given $\hat{Q}$-box determines how many times it should be differentiated. If m arrows are pointing towards it, we associate with it a factor $\frac{1}{m!}\frac{d^m\hat{Q}}{d\omega^m}$. To have this term, there must be at least m $\hat{Q}$-boxes to its right. The general expression for an n-folded $\hat{Q}$-box diagram is then

$$\hat{Q}'\int\hat{Q}\int\hat{Q} \cdots \int\hat{Q} = \sum_{m_1,m_2,\ldots,m_n}{}' \frac{1}{m_1!}\frac{d^{m_1}\hat{Q}'}{d\omega^{m_1}} P \frac{1}{m_2!}\frac{d^{m_2}\hat{Q}}{d\omega^{m_2}} P \cdots \frac{1}{m_n!}\frac{d^{m_n}\hat{Q}}{d\omega^{m_n}} P\hat{Q} , \tag{73}$$

where Σ' means that the summation satisfies the constraints

$$\begin{aligned} &m_1 + m_2 + \cdots + m_n = n , \\ &m_1 \geq 1 , \\ &m_2, m_3, \ldots, m_n \geq 0 , \\ &m_k \leq n - k + 1 . \end{aligned} \tag{73.1}$$

The last restriction is simply due to the fact that there are only n-k+1 boxes to the right of the kth box. Thus, it can at most be differentiated n-k+1 times. On the r.h.s. of eq. (73) we have inserted P-space projection operators between the $\hat{Q}$-box terms to emphasize that all P-space intermediate states are summed over. Furthermore, it is understood that all the $\hat{Q}$-boxes and $\hat{Q}$-box derivatives are evaluated at the degenerate model-space energy $\omega = W_P$.

Using the results derived above for generalized folded $\hat{Q}$-box diagrams, we can obtain a convenient prescription for calculating the effective interaction as given by eq. (49). From the relation (50) between $\hat{Q}'$ and $\hat{Q}$ it is obvious that $\frac{d\hat{Q}'}{d\omega} = \frac{d\hat{Q}}{d\omega}$, since the difference $\hat{Q}'-\hat{Q}$ does not depend on the P-space energy. Since $\hat{Q}'$ only appears in the folded terms of eq. (49) and thus is differentiated at least once, we may replace $\hat{Q}'$ by $\hat{Q}$ when the folded terms are evaluated according to eq. (73). We can then write the expansion of v_{eff} as

$$v_{eff} = F_o + F_1 + F_3 + \cdots , \tag{74}$$

where

$$F_o = \hat{Q} , \tag{74.1}$$

$$F_1 = \frac{d\hat{Q}}{d\omega} P\hat{Q} , \tag{74.2}$$

$$F_2 = \frac{1}{2!}\frac{d^2\hat{Q}}{d\omega^2} P\hat{Q}P\hat{Q} + \frac{d\hat{Q}}{d\omega} P \frac{d\hat{Q}}{d\omega} P\hat{Q} , \tag{74.3}$$

$$\begin{aligned} F_3 = {} & \frac{1}{3!}\frac{d^3\hat{Q}}{d\omega^3} P\hat{Q}P\hat{Q}P\hat{Q} + \frac{1}{2!}\frac{d^2\hat{Q}}{d\omega^2} P \frac{d\hat{Q}}{d\omega} P\hat{Q}P\hat{Q} + \frac{1}{2!}\frac{d^2\hat{Q}}{d\omega^2} P\hat{Q}P \frac{d\hat{Q}}{d\omega} P\hat{Q} \\ & + \frac{d\hat{Q}}{d\omega} P \frac{1}{2!}\frac{d^2\hat{Q}}{d\omega^2} P\hat{Q}P\hat{Q} + \frac{d\hat{Q}}{d\omega} P \frac{d\hat{Q}}{d\omega} P \frac{d\hat{Q}}{d\omega} P\hat{Q} , \end{aligned} \tag{74.4}$$

etc.

Here, all the $\hat{Q}$-boxes and $\hat{Q}$-box derivatives are evaluated at the unperturbed P-space energy $\omega = W_P$. It may be a useful exercise to verify that F_4 contains fourteen terms.

<u>5.5. Connection with the Goldstone (non-degenerate) expansion.</u> The folded-diagram theory described so far is designed to treat physical problems with a degenerate P-space, for example the sd-shell calculation of the 0^+ states in ^{18}O discussed in sect. 1.

The folded-diagram theory must still hold for problems with a <u>non-degenerate</u> P-space. Thus, it must reduce to the Goldstone theory, which is applicable only to systems with a non-degenerate P-space. We shall discuss the correspondence between the folded-diagram and Goldstone theories by way of examples.

Consider the calculation of the ground-state energy of a system of four particles in a two-level model [such as the Lipkin model[35]], where each of the levels is four-fold degenerate:

$$\begin{array}{ll} \sigma = +1 & \underline{\;p_1 \;\; p_2 \;\; p_3 \;\; p_4\;} \\ & \qquad\qquad\qquad\Big\}\ \text{4 particles} \\ \sigma = -1 & \overline{\;h_1 \;\; h_2 \;\; h_3 \;\; h_4\;} \end{array}$$

An equivalent example would be, say, the calculation of the ground-state energy of ^4He using a $(0s)^4$ P-space.

In the <u>Goldstone</u> theory we take the unperturbed ground state as

$$|c\rangle = a^{\dagger}_{h_1} a^{\dagger}_{h_2} a^{\dagger}_{h_3} a^{\dagger}_{h_4} |0\rangle , \tag{75}$$

and obtain for the ground-state energy shift the well-known result

$$\Delta E_o = \langle c|H_1(t=0)U(0,-\infty)|c\rangle_L$$

 (75.1)

On the other hand, in the folded-diagram theory we shall treat both p_i and h_i $(i = 1, \ldots, 4)$ as particle states. The one-dimensional secular equation is

$$\{H_o(V) + v_{eff}(V)\}P\Psi = E_o P\Psi , \tag{76}$$

where

$$|P\Psi\rangle = |c\rangle = a^\dagger_{h_1} a^\dagger_{h_2} a^\dagger_{h_3} a^\dagger_{h_4} |0\rangle . \tag{76.1}$$

Note that all four particles are valence particles, thus we do not have any core. We readily obtain for the ground-state energy shift

$$\Delta E_o = E_o - \langle c|H_o(V)|c\rangle = \langle c|v_{eff}(V)|c\rangle$$

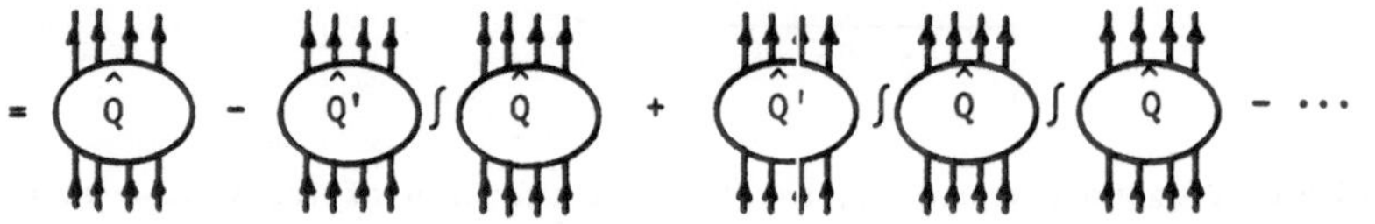 (76.2)

We now proceed to discuss the correspondence between the diagrams contained in eqs. (76.2) and (75.1). First, we consider typical diagrams contained in eq. (76.2). A few examples are shown in fig. 24. Diagrams A1, A2 and A3 contribute to $\hat{Q}$, while diagrams A4 and A5 contribute to $\hat{Q}'\int\hat{Q}$. Clearly, the active lines ($|$) represent h-states and the passive lines ($\updownarrow$) p-states.

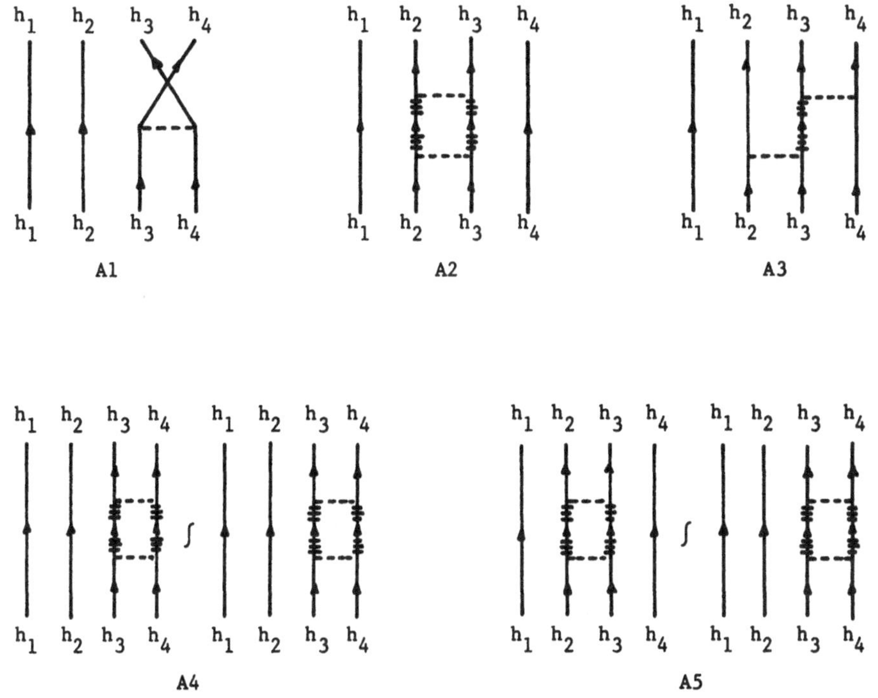

Fig. 24. Typical diagrams contained in the folded-diagram expansion of ΔE_o .

The Goldstone counterparts of the diagrams A1-A5 shown in fig. 24 are readily obtained by closing all the h-lines onto themselves. This is equivalent to factorizing out a common core propagator $\underset{h_1}{|}\ \underset{h_2}{|}\ \underset{h_3}{|}\ \underset{h_4}{|}$. The Goldstone diagrams obtained are shown in fig. 25.

The correspondence between the diagrams A1-A3 and B1-B3 is straightforward. There are two Goldstone diagrams (B4 and B4') corresponding to the generalized folded diagram A4 (see, however, discussion below). In order to facilitate the transition from A4 to B4 and B4' we have gone via the intermediate forms C4 and C4' in fig. 25. Similarly, it is realized that the generalized folded diagram A5 gives rise to the familiar BBP diagrams[36] B5 and B5' in the Goldstone picture.

A subtle, yet important point should now be emphasized. Using the diagram rules to be discussed in the following section, one can readily verify that the contribution to the energy shift ΔE_o from diagram A5 and

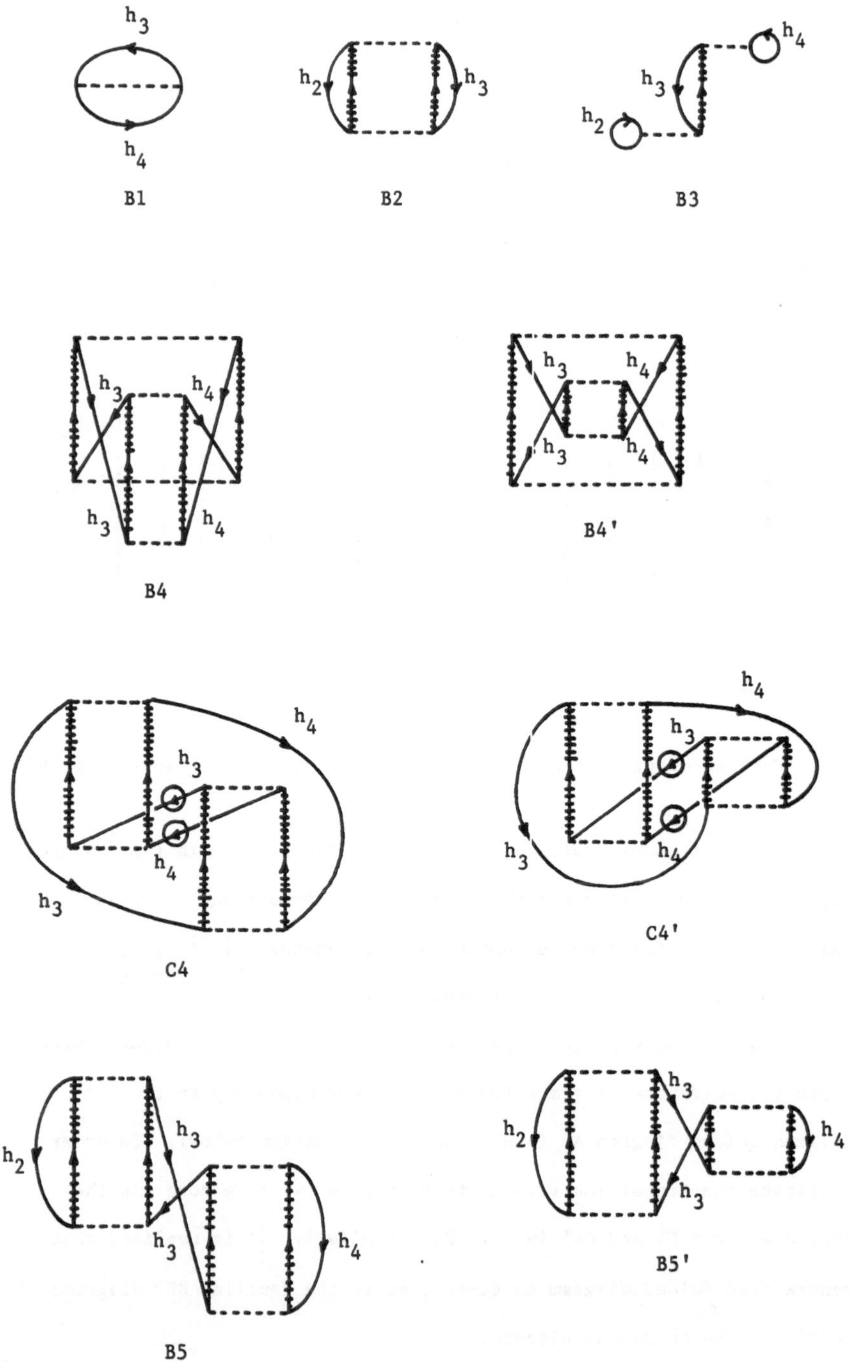

Fig. 25. Goldstone representation of the Feynman diagrams shown in fig. 24. Diagrams A1-A3 in fig. 24 give rise to B1-B3, respectively, while A4 gives rise to B4 and B4' and A5 gives rise to B5 and B5'. Diagrams C4 and C4' are topologically equivalent to B4 and B4', respectively.

from diagrams B5 and B5' are equal. But the contribution to ΔE_0 from A4 is not equal to that from B4 and B4'. Why is this so? As discussed by Ellis and Osnes [see Appendix A of ref.[12)]], diagram A4 is not the only folded diagram which can contribute to diagrams B4 and B4'. Exchange diagrams of A5, corresponding to permutations of its external hole lines, can also contribute to B4 and B4'. Equivalence will be restored when these are also included.

Clearly, the folded-diagram and Goldstone expansions for ΔE_0 must be equivalent. This allows us to regroup the Goldstone diagrams according to the number of folds. It would be of interest to examine if this regrouping has any effect on the convergence of the Goldstone expansion. This might for instance be tested by using the non-degenerate folded-diagram expansion to calculate the binding energy of nuclear matter.

As a final remark on the equivalence between the non-degenerate folded-diagram and Goldstone expansions we mention that the cancellation of valence-disconnected diagrams in the folded-diagram series corresponds to the cancellation of unlinked diagrams in the Goldstone expansion. Furthermore, in the folded-diagram series we do not have diagrams like the one shown in fig. 26a, as discussed for fig. 17. This is consistent with the fact that the corresponding Goldstone diagram shown in fig. 26b is unlinked.

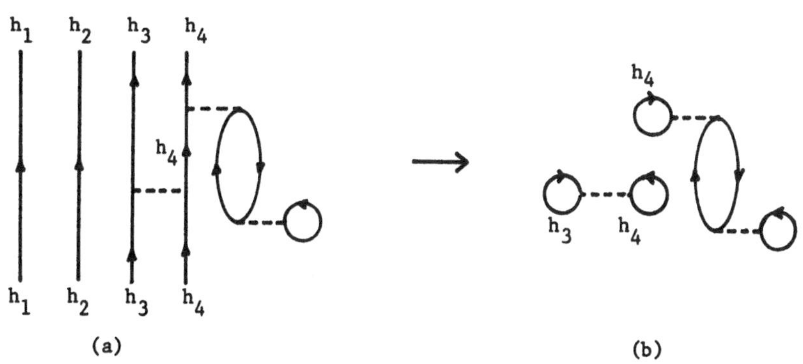

Fig. 26. Forbidden diagram in the folded-diagram expansion (a) which is unlinked in the Goldstone representation (b).

6. Diagram rules

Diagrams of the effective interaction v_{eff} can be calculated in the following two schemes:

(i) The time-ordered non-folded and folded diagrams are calculated individually.

(ii) The non-folded diagrams are calculated first and summed to give the $\hat{Q}$-box. Then, generalized folded diagrams are evaluated in terms of the $\hat{Q}$-box and its energy derivatives.

The diagram rules for the above calculations are rather straightforward, being very similar to the rules for calculating the ground-state energy shift using Goldstone or Hugenholtz diagrams. It is convenient to use antisymmetrized Hugenholtz vertices

$$V = \frac{1}{4} \sum_{ijk\ell} v_{ij,k\ell}\, a_i^\dagger a_j^\dagger a_\ell a_k , \tag{77}$$

where $v_{ij,k\ell}$ is the antisymmetrized matrix element of V

$$v_{ij,k\ell} = \langle 0|a_j a_i V a_k^\dagger a_\ell^\dagger|0\rangle = -v_{ij,\ell k} = -v_{ji,k\ell} . \tag{77.1}$$

Now, consider the matrix element

$$I_{\alpha\beta\gamma\delta} \equiv \langle c|a_\beta a_\alpha H_1(t=0)U(0,-\infty)a_\gamma^\dagger a_\delta^\dagger|c\rangle_V , \tag{78}$$

where $H_1 = V - U$ (U being an auxiliary one-body potential) and the subscript V denotes that all the vertices of H_1 are linked to at least one valence line. The time evolution operator $U(0,-\infty)$ is expressed in the interaction picture and is given by eq. (5), which we repeat here for convenience

$$U(0,-\infty) = \lim_{t'\to -\infty(\varepsilon)} \sum_{n=0}^{\infty} \left(\frac{-i}{\hbar}\right)^n \int_{t'}^{0} dt_1 \int_{t'}^{t_1} dt_2 \cdots \int_{t'}^{t_{n-1}} dt_n\, H_1(t_1)H_1(t_2) \cdots H_1(t_n) . \tag{78.1}$$

From what we have learned in the previous sections, all the terms contributing to the matrix element $\langle\alpha\beta|v_{eff}|\gamma\delta\rangle$ originate from $I_{\alpha\beta\gamma\delta}$:

(i) The non-folded terms of $\langle\alpha\beta|v_{eff}|\gamma\delta\rangle$ are just those terms of $I_{\alpha\beta\gamma\delta}$ in which all the intermediate states have at least one passive line.

(ii) The folded terms of $\langle\alpha\beta|v_{eff}|\gamma\delta\rangle$ are those terms of $I_{\alpha\beta\gamma\delta}$ in which one or more intermediate states are entirely composed of active lines, and thus have to be folded to make $I_{\alpha\beta\gamma\delta}$ finite. Note that the folded terms are at least third order in H_1 (see caption to table 1), i.e. the intermediate state between the top two vertices must be passive.

With these observations in mind it is straightforward to obtain the diagram rules for v_{eff}. It is convenient to do so by way of simple examples:

(a) $n = 0$ case: For $n = 0$ the contribution to v_{eff} from $I_{\alpha\beta\gamma\delta}$ is just

$$\langle c|a_\beta a_\alpha H_1(t=0)a_\gamma^\dagger a_\delta^\dagger|c\rangle , \tag{79}$$

where

$$H_1(t=0) = H_1 = V - U = \frac{1}{4}\sum_{ijk\ell} v_{ij,k\ell}a_i^\dagger a_j^\dagger a_\ell a_k - \sum_{ij} U_{ij}a_i^\dagger a_j . \tag{79.1}$$

We use Wick's theorem to work out the contractions in the matrix element (79); this is readily done since $v_{ij,k\ell}$ is the antisymmetrized two-body vertex. We have

$$\langle c|a_\beta a_\alpha H_1 a_\gamma^\dagger a_\delta^\dagger|c\rangle = v_{\alpha\beta,\gamma\delta} - \delta_{\beta\delta}U_{\alpha\gamma} - \delta_{\alpha\gamma}U_{\beta\delta} + \delta_{\alpha\delta}U_{\beta\gamma} + \delta_{\beta\gamma}U_{\alpha\delta} \tag{79.2a}$$

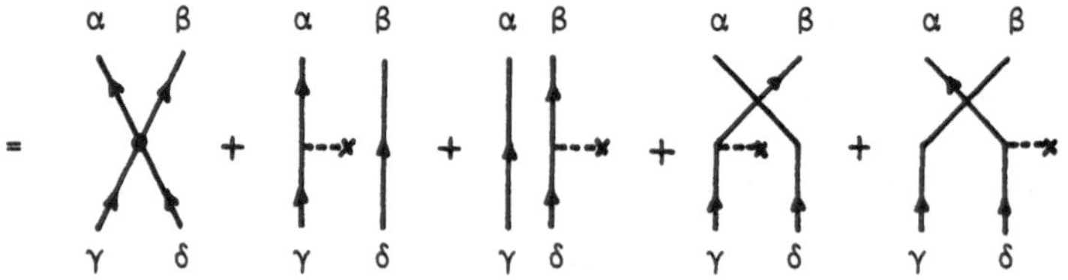

(79.2b)

where we have adopted the following rules for the diagrammatic representation of the various contractions:

(i) The vertices $v_{ij,k\ell}$ and $-U_{ij}$ are respectively represented by

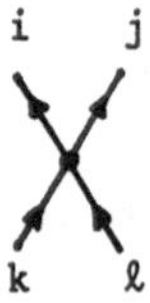

and

or

(80)

(ii) The bra and ket vectors are represented by collections of fermion lines which, if contracted with operators in H_1, are leaving and entering the vertex, respectively. A contraction between an operator in the bra vector and an operator in the ket vector is represented by an undisturbed fermion line all the way through from the bottom to the top of the diagram. Note also that we have adopted a specific order for the labelling of external fermion lines. For the fermion lines in the bra vector $\langle c|a_{\alpha_n} a_{\alpha_{n-1}} \cdots a_{\alpha_1}$ and the ket vector $a^\dagger_{\beta_1} a^\dagger_{\beta_2} \cdots a^\dagger_{\beta_n}|c\rangle$ we shall always use the order of labelling shown in fig. 27. When this convention is adopted, we shall have a minus sign for each crossing of a pair of external lines. For example, the fourth term on the r.h.s. of eq. (79.2b) corresponds to the contraction

$$\sum_{ij} (-U_{ij})\, a_\beta a_\alpha a_i^\dagger a_j a_\gamma^\dagger a_\delta^\dagger = U_{\beta\gamma}\delta_{\alpha\delta}\,, \qquad (81)$$

where the extra minus sign on the r.h.s. obviously is due to the crossing of external lines.

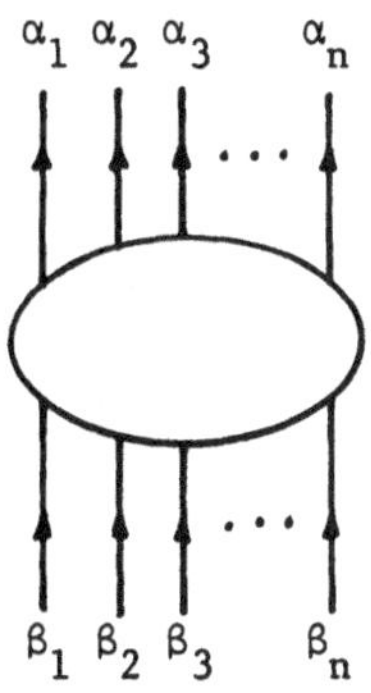

Fig. 27. Order of labelling of external fermion labels.

(b) n = 1 case: There are no folded diagrams in second order of the perturbation H_1. Thus, the contribution to v_{eff} is just the n = 1 part of $I_{\alpha\beta\gamma\delta}$, as given by eqs. (78) and (78.1)

$$<c|a_\beta a_\alpha H_1(t=0)(\frac{-i}{\hbar})\int_{-\infty}^{0} dt_1 H_1(t_1)a_\gamma^\dagger a_\delta^\dagger|c>_V . \quad (82)$$

To evaluate this term we have to make all possible contractions on the operator product

$$a_\beta a_\alpha \sum_{1,2,3,4} \frac{1}{4} v_{12,34}(a_1^\dagger a_2^\dagger a_4 a_3)_{t=0} \sum_{5,6,7,8} \frac{1}{4} v_{56,78}(a_5^\dagger a_6^\dagger a_8 a_7)_{t=t_1} a_\gamma^\dagger a_\delta^\dagger . \quad (82.1)$$

Typical diagrams contained in the n = 1 part of $I_{\alpha\beta\gamma\delta}$ are shown in fig. 28. Again, we are using antisymmetrized Hugenholtz vertices. Furthermore, we are drawing only topologically distinct diagrams. Thus, each of the diagrams shown in fig. 28 represents a class of topologically equivalent† diagrams, as explained in detail below.

† Two Hugenholtz diagrams are said to be topologically equivalent if they can be transformed into each other by deforming (twisting, bending, stretching, etc.) the fermion lines (without affecting the relative ordering of the vertices and the ordering of the external fermion labels). Diagrams which are not topologically equivalent are said to be topologically distinct.

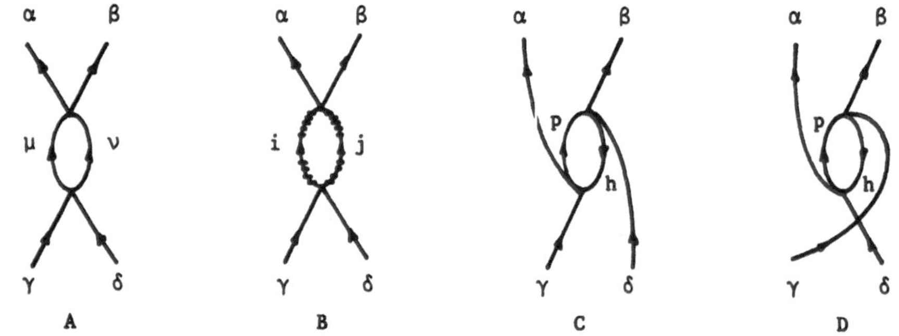

Fig. 28. Diagrams representing various sets of contractions in the $n = 1$ part of $I_{\alpha\beta\gamma\delta}$.

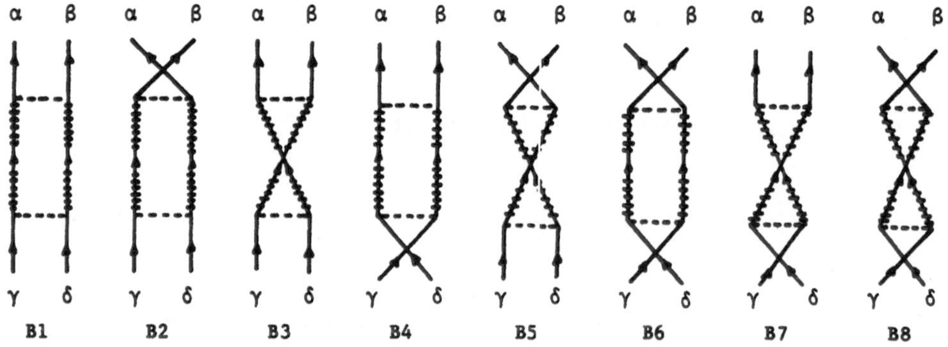

Fig. 29. Class of topologically equivalent contractions corresponding to diagram B in fig. 28. For clarity the dot vertex of fig. 28 (·) has been stretched into a dashed line.

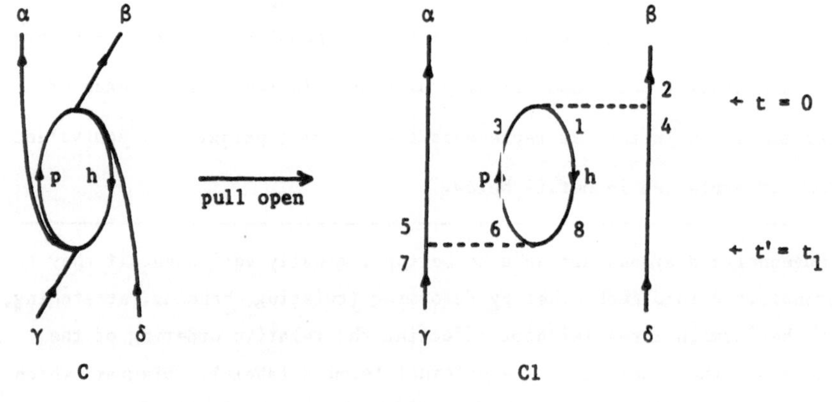

Fig. 30. One of 16 equivalent sets of contractions corresponding to diagram C

First, we note that diagram A in fig. 28 is not allowed. This is because there is no passive line in the intermediate state. Then, we observe that diagram B corresponds to the sum of eight different sets of contractions, which are shown in fig. 29. Each of the diagrams in fig. 29 corresponds to a specific set of contractions, for example

$$a_\beta a_\alpha \, (a_1^\dagger a_2^\dagger a_4 a_3)_{t=0} \, (a_5^\dagger a_6^\dagger a_8 a_7)_{t_1} \, a_\gamma^\dagger a_\delta^\dagger \quad \rightarrow \quad B5 \, , \tag{82.2}$$

where we see that there are two crossings on the l.h.s. Since we use antisymmetrized vertices, all the eight terms B1-B8 are equivalent. Thus, it does not matter which of the eight terms is used for calculation.

We have thus arrived at the following procedure. First, we draw topologically distinct diagrams in terms of dot vertices. Then, we may stretch the dot vertex into a line vertex in any way we like. For example, using the set of contractions represented by diagram B1 in fig. 29, we have

$$B = 8 \times B1 = 8 \times \left(\frac{1}{4}\right)^2 \sum_{\substack{ij \\ \text{passive}}} v_{\alpha\beta,ij} \, v_{ij,\gamma\delta} \left(\frac{-i}{\hbar}\right) \int_{-\infty(\varepsilon)}^{0} dt_1 \, e^{-i(\varepsilon_\gamma+\varepsilon_\delta-\varepsilon_i-\varepsilon_j)t_1/\hbar}$$

$$= \frac{1}{2} \sum_{\substack{ij \\ \text{passive}}} \frac{v_{\alpha\beta,ij} \, v_{ij,\gamma\delta}}{\varepsilon_\gamma+\varepsilon_\delta-\varepsilon_i-\varepsilon_j} \, . \tag{82.3}$$

The factor 1/2 is the familiar factor associated with the pair i,j of equivalent lines (see below).

Consider now diagram C in fig. 28, which represents the sum of 16 equivalent terms, each corresponding to a specific set of contractions (or a specific stretching of the dot vertices). We shall use this diagram to work out the rules for hole lines and closed loops. Thus, let us evaluate diagram C in detail. One way of pulling open diagram C is shown in fig. 30.

The resulting diagram is denoted by C1. Recall that the contractions corresponding to this diagram are made on the operator product (82.1). Since all the operators in this product are contracted, we may first regroup the vertex operators as follows:

$$a_1^\dagger a_2^\dagger a_4 a_3 \quad \rightarrow \quad \underbrace{(a_1^\dagger a_3)}_{\text{left}}\underbrace{(a_2^\dagger a_4)}_{\text{right}} , \tag{82.4}$$

$$a_5^\dagger a_6^\dagger a_8 a_7 \quad \rightarrow \quad \underbrace{(a_5^\dagger a_7)}_{\text{left}}\underbrace{(a_6^\dagger a_8)}_{\text{right}} . \tag{82.5}$$

Then, we may move the bracketed pairs of operators where we want. These operations obviously give <u>no</u> sign changes. Furthermore, we may regroup the external operators a_α, a_β, $a_\gamma^\dagger$, $a_\delta^\dagger$ according to which continuous fermion line they belong to:

$$a_\beta a_\alpha a_\gamma^\dagger a_\delta^\dagger \quad \rightarrow \quad \underbrace{(a_\alpha a_\gamma^\dagger)}_{\text{left line}}\underbrace{(a_\beta a_\delta^\dagger)}_{\text{right line}} . \tag{82.6}$$

Hence, the contractions corresponding to diagram C1 in fig. 30 can be regrouped, without introducing any extra sign factors, in the following way:

$$[(a_\alpha a_\gamma^\dagger)(a_5^\dagger a_7)]\ [(a_1^\dagger a_3)(a_6^\dagger a_8)]\ [(a_\beta a_\delta^\dagger)(a_2^\dagger a_4)]$$

$$= \underbrace{[a_\alpha (a_5^\dagger a_7) a_\gamma^\dagger]}_{\text{left valence line}} \times \underbrace{[(-) a_3 a_6^\dagger a_8 a_1^\dagger]}_{\text{middle fermion loop}} \times \underbrace{[a_\beta (a_2^\dagger a_4) a_\delta^\dagger]}_{\text{right valence line}} . \tag{82.7}$$

Here, the middle minus sign is due to the odd permutation necessary to bring the operators corresponding to the closed fermion loop into the standard order

$\overline{a_\alpha a^\dagger_\beta}$ chosen for contracted operators (i.e. a to the left of $a^\dagger$). Furthermore, the standard order contraction $\overline{a_8 a^\dagger_1}$ for the hole line implicitly contains a minus sign. Thus, it is clear that the sign factors for hole lines and closed loops are

$$(-)^{n_h+n_\ell} \ , \quad \begin{cases} n_h = \text{number of hole lines} \\ n_\ell = \text{number of closed loops} . \end{cases} \tag{82.8}$$

By evaluating the time integrals corresponding to diagram C1, as we did for diagram B in eq. (82.3), we obtain for diagram C in fig. 28

$$\begin{aligned} C &= 16 \times C1 = 16 \times (-)^{n_h+n_\ell} (\tfrac{1}{4})^2 \sum_{\substack{p>h_F \\ h\le h_F}} \frac{v_{h\beta,p\delta}\, v_{\alpha p,\gamma h}}{(\varepsilon_\gamma+\varepsilon_\delta)-(\varepsilon_\alpha+\varepsilon_p-\varepsilon_h+\varepsilon_\delta)} \\ &= \sum_{\substack{p>h_F \\ h\le h_F}} \frac{v_{h\beta,p\delta}\, v_{\alpha p,\gamma h}}{(\varepsilon_\gamma+\varepsilon_\delta)-(\varepsilon_\alpha+\varepsilon_p-\varepsilon_h+\varepsilon_\delta)} \ , \end{aligned} \tag{82.9}$$

where we have used $n_h = 1$ and $n_\ell = 1$. Since there are no pairs of equivalent fermion lines (see below), the factors 1/4 associated with the v vertices are exactly cancelled away. Note that the summations over p and h are unrestricted. Thus, p may be equal to α or δ. Also, α may be equal to δ. This should cause no worry, since in the linked-diagram perturbation expansion we always violate the Pauli exclusion principle in the intermediate states of the individual linked diagrams[37].

The diagram rules for the valence-linked v_{eff} differ from the Goldstone (and Hugenholtz) diagram rules for the ground-state energy shift because of the presence of external lines. In Goldstone diagrams there are no external lines, and we have vacuum both at $t' = -\infty$ and at $t = 0$.

The additional rules due to external lines are:

(i) There is a factor $(-)^{n_c}$ due to the crossing of n_c pairs of external lines.

(ii) There is a factor $(-)^{n_{exh}}$ due to the presence of n_{exh} continuous external hole lines.

The former rule was briefly discussed above [see eq. (81)]; here we give a further example. Consider diagram D in fig. 28 which may be pulled open as shown in fig. 31. By regrouping the operators a_α, a_β, $a_\gamma^\dagger$, $a_\delta^\dagger$ according to which external line they belong, we have

$$a_\beta a_\alpha a_\gamma^\dagger a_\delta^\dagger \quad \rightarrow \quad \underbrace{-(a_\alpha a_\delta^\dagger)}_{\substack{\text{for} \\ \alpha 57\delta \text{ line}}} \underbrace{(a_\beta a_\gamma^\dagger)}_{\substack{\text{for} \\ \beta 24\gamma \text{ line}}} , \qquad (82.10)$$

Since the γ and δ lines are crossing each other, an odd permutation is needed to bring together the external fermion operators belonging to the same continuous external lines. Thus, the set of contractions corresponding to diagram D1 in fig. 31 is

$$(-)^{n_c}[a_\alpha(a_5^\dagger a_7)a_\delta^\dagger][(-)^{n_\ell} a_3(a_6^\dagger a_8)a_1^\dagger][a_\beta(a_2^\dagger a_4)a_\gamma^\dagger] , \qquad (82.11)$$

where $n_c = 1$ and $n_\ell = 1$.

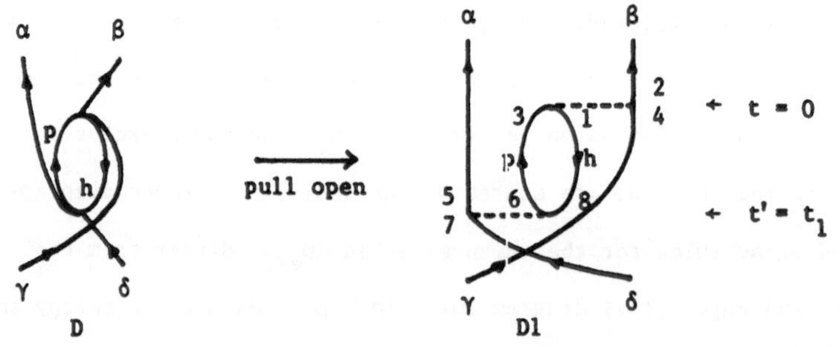

Fig. 31. One set of contractions corresponding to diagram D in fig. 28.

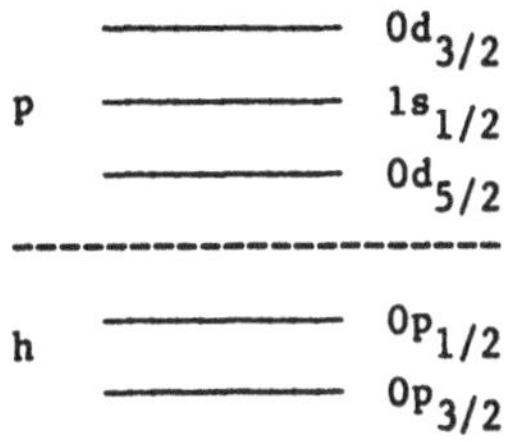

Fig. 32. Single-particle orbits for particle-hole description of ^{16}O.

We then discuss the phase factor associated with continuous external hole lines. Although we have not explicitly worked out v_{eff} for valence holes, it can be obtained by a straightforward generalization of v_{eff} for valence particles. Consider for example the low-energy excitations of ^{16}O using a P-space defined by

$$P = \sum_{ph} |1p\text{-}1h\rangle\langle 1p\text{-}1h| , \tag{83}$$

$$|1p\text{-}1h\rangle \equiv a_p^{\dagger} a_h |c\rangle . \tag{83.1}$$

Here, p and h are thought to run over the single-particle orbits shown in fig. 32. For this P-space, the effective interaction is just

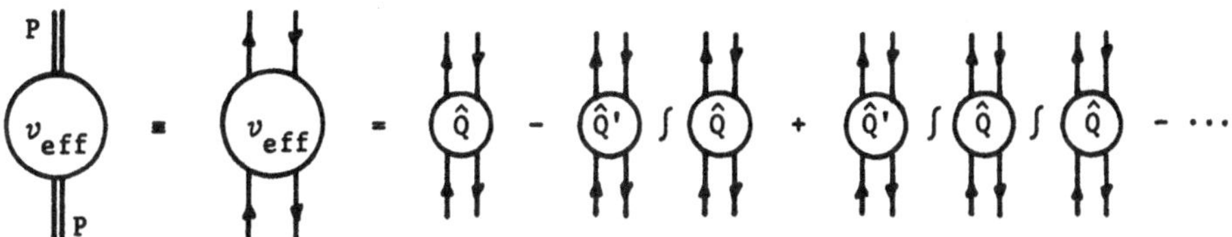

(83.2)

Note that the incoming and outgoing active lines on the far l.h.s. are each composed of one particle line and one hole line. Typical diagrams in the $\hat{Q}$-box of eq. (83.2) are shown in fig. 33. Just as for the valence-particle diagrams considered above these diagrams originate from the $n = 1$ part of the valence-linked matrix element

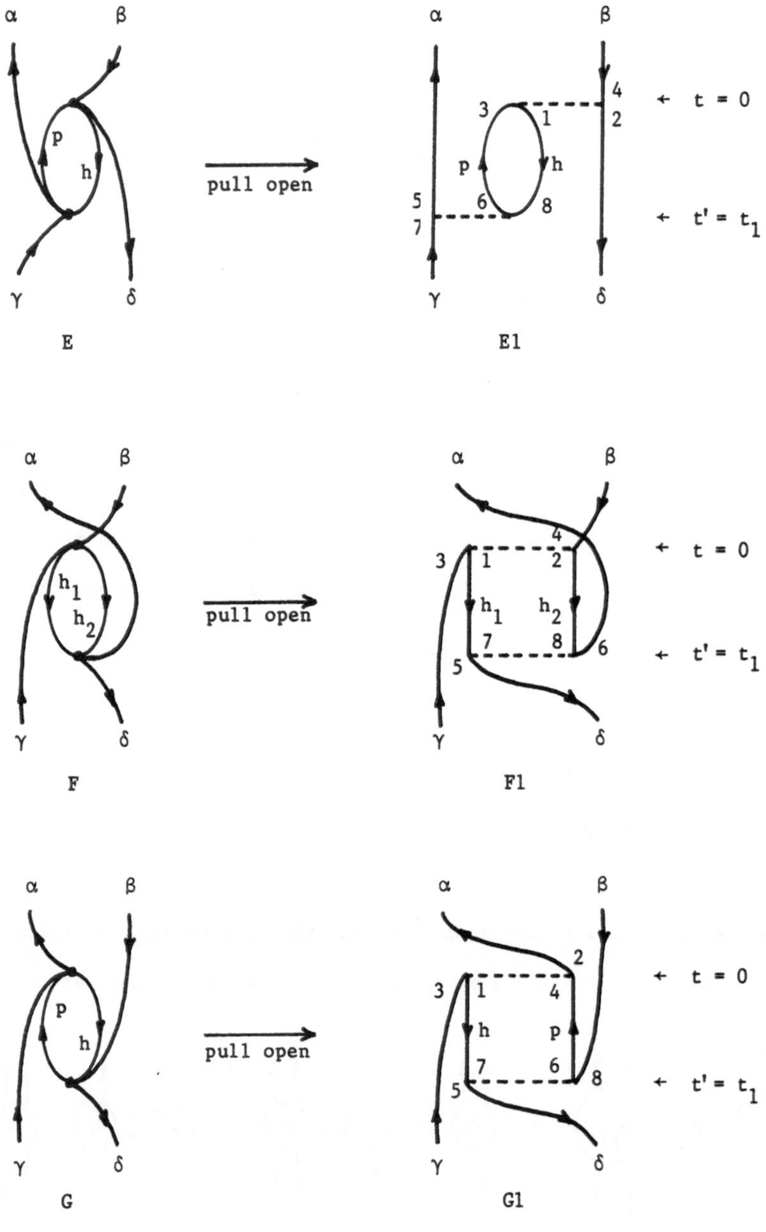

Fig. 33. Diagrams contributing to the effective particle-hole interaction.

$$I^{(1p-1h)}_{\alpha\beta\gamma\delta} = \langle c|a^{\dagger}_{\beta}a_{\alpha}H_1(t=0)U(0,-\infty)a^{\dagger}_{\gamma}a_{\delta}|c\rangle\,, \tag{83.3}$$

where $U(0,-\infty)$ is given by eq. (78.1). The contractions of diagram E1 in fig. 33 are clearly composed of three parts:

$$\text{particle line} \quad (\alpha 57\gamma) \rightarrow a_{\alpha}(a^{\dagger}_{5}a_{7})a^{\dagger}_{\gamma}\,, \tag{83.4a}$$

$$\text{closed loop} \quad (1368) \rightarrow (a^{\dagger}_{1}a_{3})(a^{\dagger}_{6}a_{8}) = (-)^{n_{\ell}}a_{3}(a^{\dagger}_{6}a_{8})a^{\dagger}_{1}\,, \quad n_{\ell} = 1, \tag{83.4b}$$

$$\text{hole line} \quad (\beta 42\delta) \rightarrow a^{\dagger}_{\beta}(a^{\dagger}_{2}a_{4})a_{\delta} = (-)^{n_{exh}}a_{\delta}(a^{\dagger}_{2}a_{4})a^{\dagger}_{\beta}\,, \quad n_{exh} = 1\,. \tag{83.4c}$$

In eq. (83.4c) it was necessary to interchange $a^{\dagger}_{\beta}$ and a_{δ} to have all the contractions in the adopted standard order (i.e. a to the left of $a^{\dagger}$). This interchange, which gives rise to an odd permutation, is attributed to the line $(\beta 42\delta)$ being a continuous external hole line.† Similarly, we have for the line $(\beta 4286\alpha)$ of diagram F1

$$a^{\dagger}_{\beta}(a^{\dagger}_{2}a_{4})(a^{\dagger}_{6}a_{8})a_{\alpha} = (-)^{n_{exh}}a_{\alpha}(a^{\dagger}_{6}a_{8})(a^{\dagger}_{2}a_{4})a^{\dagger}_{\beta}\,, \quad n_{exh} = 1\,, \tag{83.5}$$

where again we just need to interchange a_{α} and $a^{\dagger}_{\beta}$ to have all the contracted pairs in standard order.

Note that there is <u>no crossing</u> of external lines in diagram F1. To be more precise, let us define an external line as a fermion line which enters

---

† Note that this sign rule depends on having the particle operators in front of the hole operators in the ket vector, i.e. $a^{\dagger}_{p}a_{h}|c\rangle$ rather than $a_{h}a^{\dagger}_{p}|c\rangle$. If the latter ordering were used, we would have an extra minus sign for each external hole line, which is not convenient.

the "central piece" of the diagram (i.e. the one involving the vertices and the intermediate states) from $t > 0$ or $t' = -\infty$ and then leaves towards $t > 0$ or $t' = -\infty$. Thus, diagrams E1, F1 and G1 in fig. 33 each have two external lines. If an external line intersects a different external line, we have crossing of external lines. Hence, there is no crossing of external lines in diagram F1, since the lines α and β actually are pieces of the same external line.

We are now ready to write down the rules for calculating non-folded diagrams (i.e. those of the first $\hat{Q}$-box term) with external lines:

(1) Draw the external lines at $t > 0$ (upper boundary of the diagram) and $t' = -\infty$ (lower boundary of the diagram). Label the external lines in consistency with the order of creation and destruction operators used in the bra and ket vectors (see fig. 27). Note that the labelling and ordering of the external lines must not be altered.

(2) For a diagram with n H_1 interactions [i.e. belonging to the (n-1)-th term in the expansion of $U(0,-\infty)$, eq. (78.1)], draw n vertices located at the times 0, t_1, t_2, ..., t_{n-1} ordered as $0 > t_1 > t_2 > \ldots > t_{n-1}$. Each of the vertices can either be a Hugenholtz antisymmetrized v vertex (denoted by a dot $\cdot$) or a one-body $-U$ vertex (denoted by a cross $\times$).

(3) Draw topologically distinct diagrams by linking up the external lines and vertices with particle and hole lines. Note that each vertex must be valence-linked, i.e. linked (directly or indirectly) to at least one valence line. By topologically distinct diagrams we mean diagrams which are not topologically equivalent. Two diagrams are said to be topologically equivalent if they can be made identical by deformation of fermion lines under the restrictions that

(i) the time ordering of the vertices is not altered,

(ii) particle lines remain particle lines and hole lines remain hole lines, and

(iii) the ordering of external lines at $t > 0$ and $t' = -\infty$ is not altered.

According to these criteria diagrams A and B in fig. 34 are topologically equivalent, whereas diagrams A and C are topologically distinct. Furthermore, diagrams D and E are topologically equivalent and diagrams D and F topologically distinct (the lines γ and δ are interchanged).

(4) The contribution from each of the topologically distinct diagrams is calculated as follows:

(4-1) Pull open the diagram by stretching the dot vertex (·) into a dashed-line vertex (----). (It does not matter how the diagram is pulled open, since all ways of pulling will give the same final result.)

(4-2) Each v vertex gives a contribution $v_{\alpha\beta,\gamma\delta}$, where α and β refer to the lines leaving respectively the left and right end points of the vertex and γ and δ refer to the lines entering respectively the left and right end points of the vertex. (Here, it is understood that each vertex is a stretched-open dashed-line vertex.)

(4-3) Each $-U$ vertex gives a contribution $-U_{\alpha\beta}$, where α and β refer to the lines leaving and entering the vertex, respectively.

(4-4) There is a factor $(-)^{n_h+n_\ell+n_c+n_{exh}}$ associated with the diagram, where n_h is the number of hole lines, n_ℓ is the number of closed fermion loops, n_c is the number of crossings of different external lines and n_{exh} is the number of continuous external hole lines.

(4-5) For each interval between two successive vertices (note that there must be at least one passive line) with particle lines $p_1, p_2, \dots, p_m$ and hole lines $h_1, h_2, \dots, h_s$, there is a factor

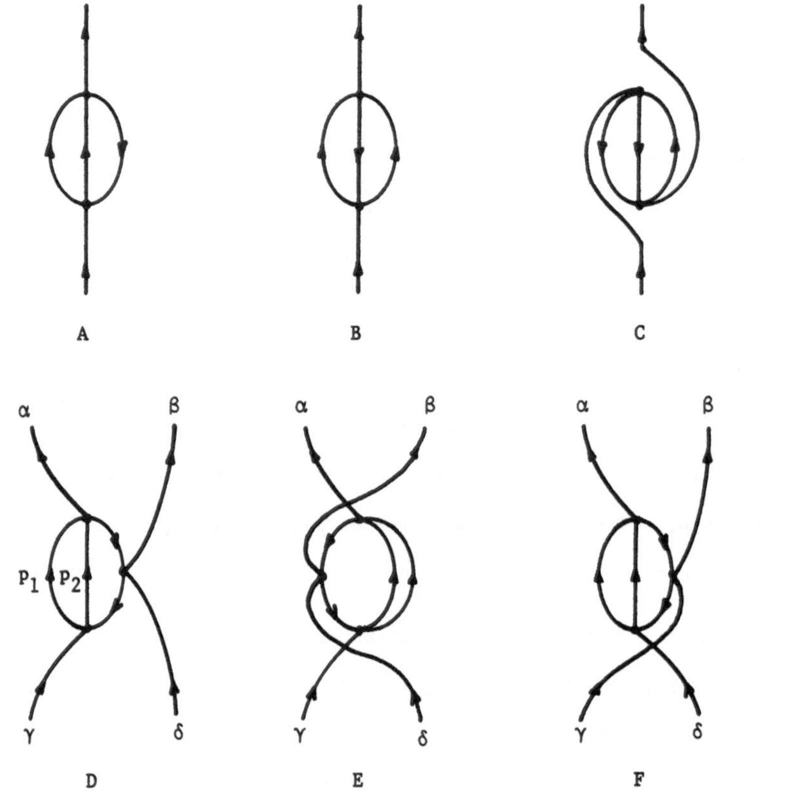

Fig. 34. Diagrams A and B are topologically equivalent, and A and C topologically distinct. Further, diagrams D and E are topologically equivalent, and D and F topologically distinct.

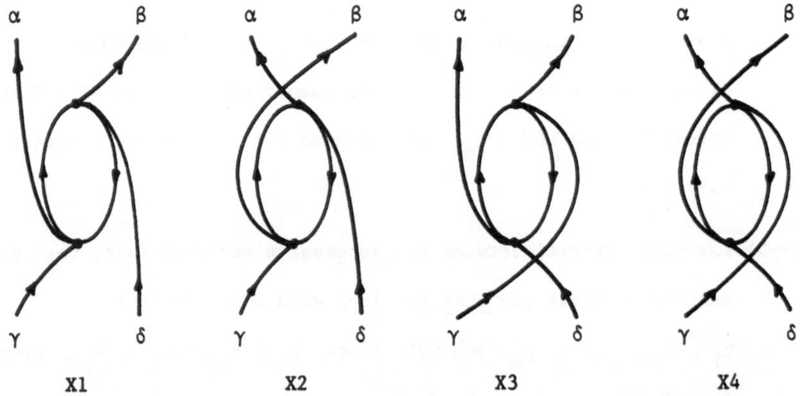

Fig. 35. Diagrams which are related to each other by interchange of

$$\left[W_P - \sum_{i=1}^{m} \varepsilon_{p_i} + \sum_{j=1}^{s} \varepsilon_{h_j}\right]^{-1},$$

where W_P is the sum of the unperturbed energies corresponding to the external lines at $t' = -\infty$.

(4-6) There is a factor $1/2^{n_{ep}}$, where n_{ep} is the number of pairs of lines which start at the same interaction, end at the same interaction and go in the same direction (so-called equivalent pairs).†

(4-7) Sum freely over the labels of the internal fermion lines (i.e. between the vertices). [For example, in diagram D of fig. 34 we sum over

$$(p_1, p_2) = (1,1),\ (1,2),\ (2,1) \text{ and } (2,2)$$

if the range of both p_1 and p_2 is from 1 to 2.]

(5) The sum of all the topologically distinct diagrams with n H_1 vertices and external lines $\alpha_1, \alpha_2, \ldots, \alpha_m$ at $t > 0$ and $\beta_1, \beta_2, \ldots, \beta_{m'}$ at $t' = -\infty$ gives the matrix element

$$\langle \alpha_1 \alpha_2 \ldots \alpha_m | \hat{Q}(W_P) | \beta_1 \beta_2 \ldots \beta_{m'} \rangle_{(n)} .$$

Here the subscript (n) signifies that the above matrix element is not the entire $\hat{Q}$-box, but just that part of the $\hat{Q}$-box which has n vertices. Thus, the matrix element of the entire $\hat{Q}$-box is

$$\langle \alpha_1 \alpha_2 \ldots \alpha_m | \hat{Q}(W_P) | \beta_1 \beta_2 \ldots \beta_{m'} \rangle$$

$$= \sum_{n=1}^{\infty} \langle \alpha_1 \alpha_2 \ldots \alpha_m | \hat{Q}(W_P) | \beta_1 \beta_2 \ldots \beta_{m'} \rangle_{(n)} . \qquad (84)$$

(6) A word of caution about interchange of external lines is in order. Consider the four topologically distinct diagrams shown in fig. 35.

† This rule follows from the fact that we use antisymmetrized vertices.

All these diagrams must be included in $\langle\alpha\beta|\hat{Q}|\gamma\delta\rangle$. But it is helpful to notice that the algebraic expressions for these four diagrams are practically identical; they are in fact related to each other by the interchange of various external-line labels and the associated phase factors. For example, if diagram X1 is given by

$$X_1 = X(\alpha\beta\gamma\delta) , \tag{85}$$

we have

$$X_2 = (-)^1 X(\beta\alpha\gamma\delta) , \tag{85.1}$$

$$X_3 = (-)^1 X(\alpha\beta\delta\gamma) , \tag{85.2}$$

$$X_4 = (-)^2 X(\beta\alpha\delta\gamma) . \tag{85.3}$$

Here, the phase factors come from $(-)^{n_c}$. Thus, to compute the four diagrams X1-X4 in fig. 35, it is only necessary to derive one of them. On the other hand, if two external lines are attached to the same v vertex, we must not include the exchange diagrams arising from the interchange of these two lines, since we use antisymmetrized v vertices (see the discussion to fig. 29 above).

We have just worked out, in rather detail, the rules for evaluating $\hat{Q}$-box diagrams (i.e. non-folded) with external valence lines. As mentioned above, these diagrams arise in the case of two valence lines from the matrix element

$$I_{\alpha\beta\gamma\delta} = \langle c|a_\beta a_\alpha H_1(t=0)U(0,-\infty)a_\gamma^\dagger a_\delta^\dagger|c\rangle_V , \tag{86}$$

first defined in eq. (78). Here, the subscript V indicates that any vertex must be linked (directly or indirectly) to at least one valence line. With little modification, these diagram rules can also be used for evaluation of the folded diagrams contained in $I_{\alpha\beta\gamma\delta}$.

As pointed out in the discussion to eqs. (78-78.1) the $\hat{Q}$-box diagrams are just those terms in $I_{\alpha\beta\gamma\delta}$ where all the intermediate states have at least one passive line. On the other hand, the folded diagrams correspond to those terms in $I_{\alpha\beta\gamma\delta}$ where one or more intermediate states entirely consist of active lines, which are then folded. (Recall, however, that the intermediate state between the top two vertices must be passive.) We use the example shown in fig. 36 to illustrate the rules for evaluation of folded diagrams.

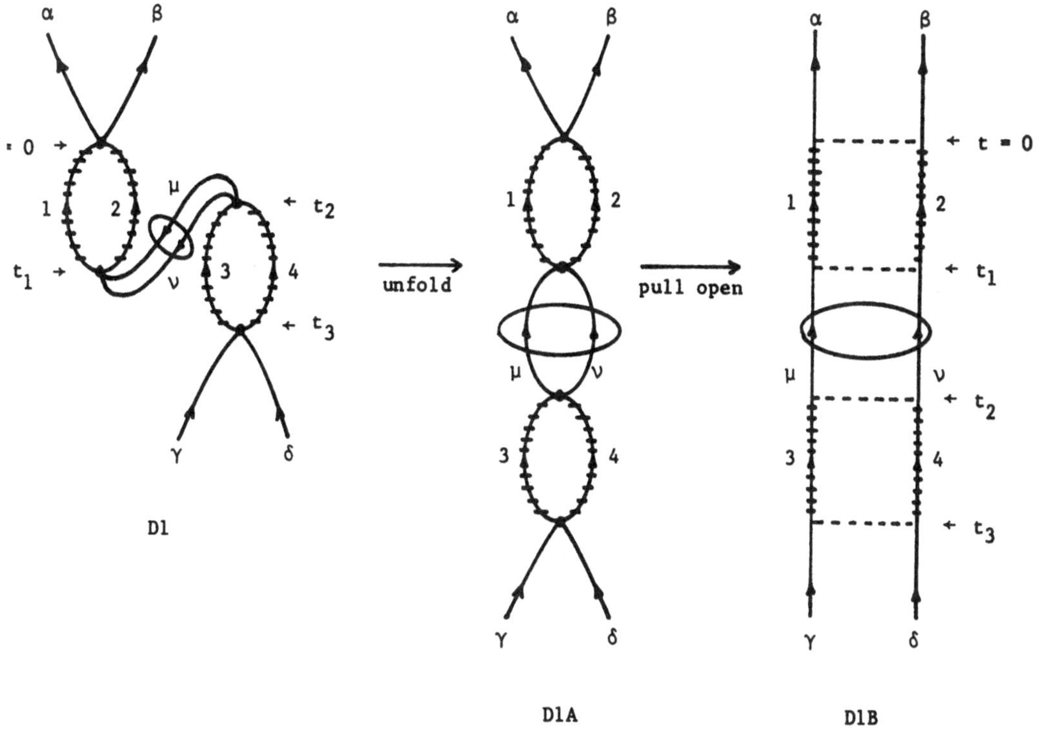

Fig. 36. Time-ordered folded diagram which is unfolded (D1A) and pulled open (D1B). The folded lines are encircled. This diagram is evaluated in eq. (87.4).

Diagrams D1 and D1A are identical in the vertices and in the labels of the propagators, but differ in the limits of the time integrations:

$$D1 \rightarrow \int_{-\infty}^{0} dt_2 \int_{-\infty}^{t_2} dt_1 \int_{-\infty}^{t_1} dt_3 , \tag{87a}$$

$$D1A \rightarrow \int_{-\infty}^{0} dt_1 \int_{-\infty}^{t_1} dt_2 \int_{-\infty}^{t_2} dt_3 . \tag{87b}$$

We can express the two diagrams as

$$D1 = N(D1)/\Delta(D1) , \tag{87.1a}$$

$$D1A = N(D1A)/\Delta(D1A) , \tag{87.1b}$$

where the Δ's are the energy denominators which are obtained by carrying out the time integrations over the propagators. Thus, $\Delta(D1A)$ is not well defined. Clearly, we have $N(D1) = N(D1A)$, and it is convenient to use diagram D1A to work out this quantity. Pulling open diagram D1A as shown in diagram D1B (note that all ways of pulling open are equivalent), we readily have

$$N(D1) = (-)^{n_h+n_\ell+n_c+n_{exh}} \left(\tfrac{1}{2}\right)^{n_{ep}} v_{\alpha\beta,12}\, v_{12,\mu\nu}\, v_{\mu\nu,34}\, v_{34,\gamma\delta} ,$$

$$n_h = n_\ell = n_c = n_{exh} = 0 \quad \text{and} \quad n_{ep} = 3 , \tag{87.2}$$

where the v's are antisymmetrized Hugenholtz vertices. We shall use diagram D1 to work out the energy denominators. We readily find, by performing the time integrations

$$\Delta(D1) = (\varepsilon_\gamma+\varepsilon_\delta-\varepsilon_1-\varepsilon_2)(\varepsilon_\gamma+\varepsilon_\delta+\varepsilon_\mu+\varepsilon_\nu-\varepsilon_1-\varepsilon_2-\varepsilon_3-\varepsilon_4)(\varepsilon_\gamma+\varepsilon_\delta-\varepsilon_3-\varepsilon_4) . \tag{87.3}$$

Thus, the value of diagram D1 is given by

$$D1 = (-)^{n_f} \sum_{\mu,\nu} \sum_{1,2} \sum_{3,4} \left(\tfrac{1}{2}\right)^3 \frac{v_{\alpha\beta,12}\, v_{12,\mu\nu}\, v_{\mu\nu,34}\, v_{34,\gamma\delta}}{(\varepsilon_\gamma+\varepsilon_\delta-\varepsilon_1-\varepsilon_2)(\varepsilon_\gamma+\varepsilon_\delta+\varepsilon_\mu+\varepsilon_\nu-\varepsilon_1-\varepsilon_2-\varepsilon_3-\varepsilon_4)(\varepsilon_\gamma+\varepsilon_\delta-\varepsilon_3-\varepsilon_4)}, \tag{87.4}$$

where $n_f = 1$ is the number of folds (see rule 4 below).

It is obvious that in order to evaluate time-ordered folded diagrams we only need to make the following modifications to the $\hat{Q}$-box diagram rules given above:

(1) It is convenient to evaluate time-ordered folded diagrams in two steps. First, the energy denominators are determined and then all the other factors.

(2) Between any two vertices we have an energy denominator

$$\left[W_P - \sum_{i=1}^{m} \varepsilon_{p_i} + \sum_{j=1}^{s} \varepsilon_{h_j} + \sum_{f=1}^{t} \varepsilon_{p_f}\right]^{-1},$$

where p_i, h_j and p_f refer to the particle, hole and folded particle lines, respectively, between these two vertices in the folded diagram. Furthermore, W_P is the sum of the unperturbed energies of the unperturbed lines at $t' = -\infty$.

(3) To obtain the other factors, one may just straighten out (unfold) the folded diagram and use the $\hat{Q}$-box diagram rules given above for the straightened diagram.

(4) There is an overall factor $(-)^{n_f}$, where n_f is the number of sets of folded lines. To determine n_f it is convenient to rearrange the the folded diagram in terms of $\hat{Q}$-boxes. For example, if the diagram is of the form

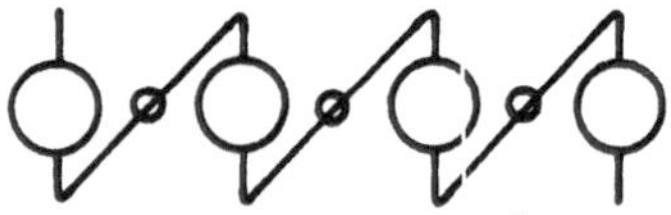

then $n_f = 3$. We emphasize that n_f is in general not equal to the number of folded single-particle lines.

(5) All intermediate-state labels are to be summed over freely, including those of folded lines. For example, in diagram D1 μ and ν are folded lines which are summed over freely within the model space. This means, for instance, that we have $(\mu,\nu) = (i,i)$, (i,j), (j,i) and (j,j) for a P-space consisting of two single-particle states i and j.

It may now be appropriate to give an example. Consider the diagram shown in fig. 37. This is a time-ordered diagram (i.e. $0 > t_3 > t_1 > t_5 > t_2 > t_4 > -\infty$) with antisymmetrized Hugenholtz vertices which are pulled open. Using the above rules we obtain for this diagram

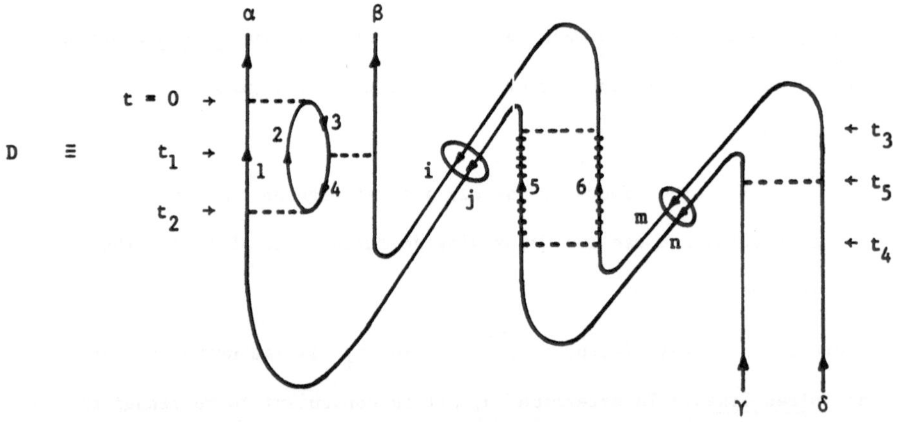

$0 > t_3 > t_1 > t_5 > t_2 > t_4 > -\infty$

Fig. 37. Twice-folded time-ordered Hugenholtz diagram which is pulled open. This diagram is evaluated in eq. (88).

$$D = (-)^{n_f} (-)^{n_h+n_\ell+n_c+n_{exh}} (\tfrac{1}{2})^{n_{ep}} \sum_{\substack{i,j,m,n \\ 1,2,3,4,5,6}}$$

$$\times \frac{v_{\alpha3,12}\, v_{4\beta,3i}\, v_{12,j4}\, v_{ji,56}\, v_{56,mn}\, v_{mn,\gamma\delta}}{(\varepsilon_\gamma+\varepsilon_\delta-\varepsilon_1-\varepsilon_2-\varepsilon_\beta+\varepsilon_3)(\varepsilon_\gamma+\varepsilon_\delta-\varepsilon_1-\varepsilon_2-\varepsilon_\beta-\varepsilon_5-\varepsilon_6+\varepsilon_3+\varepsilon_i+\varepsilon_j)(\varepsilon_\gamma+\varepsilon_\delta-\varepsilon_1-\varepsilon_2-\varepsilon_5-\varepsilon_6+\varepsilon_4+\varepsilon_j)}$$

$$\times \frac{1}{(-\varepsilon_1-\varepsilon_2-\varepsilon_5-\varepsilon_6+\varepsilon_4+\varepsilon_j+\varepsilon_m+\varepsilon_n)(-\varepsilon_5-\varepsilon_6+\varepsilon_m+\varepsilon_n)} ,$$

$$n_h = 2 , \; n_\ell = 1 , \; n_c = 0 , \; n_{exh} = 0 , \; n_{ep} = 3 \text{ and } n_f = 2 . \tag{88}$$

Here, lines 5 and 6 are passive particle lines, while lines 1 , 2 , i , j , m and n are active lines and lines 3 and 4 are hole lines. Note that in counting n_{ep} , folded lines are included. Also, all intermediate-state labels (including the folded ones) are summed over freely disregarding the Pauli exclusion principle.

It is obviously straightforward to calculate individual folded diagrams, as we have shown. It is also clear that the calculation of higher order folded diagrams becomes rather involved. Furthermore, the number of folded diagrams increases rapidly with increasing order. Because of these considerations, it may be convenient to calculate generalized folded diagrams. For diagram D of fig. 37 this means that we calculate the sum of all similar folded diagrams (i.e. having the same vertices and propagator labels) satisfying the common time constraints

$$0 > (t_1,t_2,t_3,t_4,t_5) > -\infty ,$$
$$t_3 > t_2 ,$$
$$t_5 > t_4 .$$

This sum can be expressed as products of $\hat{Q}$ and its energy derivatives when the model space is degenerate. It is simply given by eq. (74.3).

7. Application to nuclear structure calculations

In this section we discuss how to apply the folded-diagram method to actual nuclear structure calculations, using the prototype nucleus ^{18}O as an example. For convenience, let us repeat some of the basic equations derived in the preceding sections. The many-nucleon Schrödinger equation to be solved, in principle, is $H\Psi(A) = E\Psi(A)$ with $A = 18$. This is of course too complicated to be solved exactly. Thus, guided by the success of the empirical shell model, we believe that this equation can be reduced to an effective Schrödinger equation involving only two nucleons. Mathematically this is done by the folded-diagram method which we have discussed in some detail in the previous sections. In eqs. (34-34.1) we arrived at a reduced Schrödinger equation of the form

$$PH_{eff}P|b_\lambda\rangle = E_\lambda|b_\lambda\rangle \tag{89}$$

with

$$|b_\lambda\rangle = \frac{P|\Psi_\lambda\rangle}{\langle\rho_\lambda|\Psi_\lambda\rangle} . \tag{89.1}$$

Here, the parent state $|\rho_\lambda\rangle$ is of no interest and we may solve eq. (89) directly for E_λ and $|b_\lambda\rangle$. Clearly, the first step is to choose a model space P. In principle, one may choose any P-space of one's like. In practice, one should use a P-space for which the effective Hamiltonian H_{eff} can be conveniently calculated. How to choose such a P-space is a fundamental problem in nuclear many-body theory and is not yet well understood. So again, we rely on the success of the empirical shell model and take the P-space of ^{18}O to be $\{1s,0d\}^2$, i.e. the space spanned by two valence neutrons confined to the (1s0d) shell outside a closed unperturbed ^{16}O core.

The next step is the calculation of H_{eff}. First, we separate out the core energy E_C, i.e. the exact ground-state energy of ^{16}O. This leads to the model-space secular equation

$$H_{eff}P\Psi_\lambda = (E_\lambda - E_C)P\Psi_\lambda , \tag{90}$$

first given in eq. (58). Here, $H_{eff} = H_o + v_{eff}$. According to eq. (54), v_{eff} can be decomposed as follows

$$v_{eff} = v_{eff}(1) + v_{eff}(2) + \cdots + v_{eff}(N_V) , \tag{91}$$

where N_V is the number of valence nucleons. For ^{18}O, we have of course $N_V = 2$. Combining $v_{eff}(1)$ with H_o, we have for ^{18}O

$$H_{eff} = \tilde{H}_o(V) + v_{eff}(2) , \tag{92}$$

first given in eq. (58.1). Here, $\tilde{H}_o(V)$ is the one-body effective Hamiltonian given by the experimental single-particle separation energies as shown by eqs. (58.2-3). Thus, the calculation of the energies and model-space wave functions of ^{18}O is reduced to the calculation of the two-body effective interaction $v_{eff}(2)$ which, according to eq. (49), is given by

$$v_{eff}(2) = \{\hat{Q} - \hat{Q}'\int\hat{Q} + \hat{Q}'\int\hat{Q}\int\hat{Q} - \cdots\}_{2B} , \tag{93}$$

where the subscript 2B indicates that we only include the two-body connected diagrams of the $\hat{Q}$-box series. There are several ways to calculate the above $\hat{Q}$-box series. One convenient scheme is to use the energy-derivative method of sect. 5.4. Then, according to eq. (74) , we have

$$v_{eff}(2) = \{F_o + F_1 + F_2 + \cdots\}_{2B} , \tag{93.1}$$

where F_n denotes the terms of n folds as explained in eqs. (74.1-4). These are given in terms of the $\hat{Q}$-boxes and their energy derivatives. For example, $F_1 = \frac{d\hat{Q}}{d\omega} P\hat{Q}$. Thus, to calculate the F_n's , we must first know the

$\hat{Q}$-box. This is given by an infinite series composed of valence-linked irreducible diagrams.

It is worth noting that all the developments made so far, are formally exact. However, it is not possible, as far as we know, to calculate the $\hat{Q}$-box exactly. If it were, we would in fact be able to solve the Schrödinger equation exactly. Thus, we have to make approximations in evaluating the $\hat{Q}$-box. Basically, we hope that the $\hat{Q}$-box can be given accurately by certain classes of diagrams chosen by physical considerations. In fact, it is only in the calculation of the $\hat{Q}$-box that we shall make some approximation.

Then, let us discuss the evaluation of the $\hat{Q}$-box. As shown in fig. 10, the $\hat{Q}$-box diagrams are expressed in terms of vertices (}---{) of the nucleon-nucleon interaction V. (Recall that the nuclear many-body Hamiltonian is taken as $H = T + V$, where V is assumed to be a two-body nucleon-nucleon potential.) Thus, to carry out the calculation, we must first choose an appropriate nucleon-nucleon interaction V.

There exist many models of V. In fact, much work has been devoted to the determination of V [see e.g. ref.[38] and references therein]. Some well-known models of V are those of Reid[39], Holinde and collaborators[40] and Vinh Mau and collaborators[41]. These models are known respectively as the Reid, the Bonn-Jülich and the Paris potentials. Only the long-range tail of these potentials is well determined, namely it is the one-pion exchange potential. The short-range repulsive core of these potentials is most uncertain. Such a core is needed phenomenologically, to provide the change of sign in the $^1S_0$ nucleon-nucleon phase shift at a scattering energy of $E_{lab} \simeq 200$ MeV and to provide a convenient mechanism for nuclear saturation. How this repulsive core does come about is a problem which is currently under extensive study[42]. For low-energy nuclear structure calculations, the results are probably not very sensitive to the details of the repulsive core. The nuclear properties will, however, depend on the medium-range part of the nucleon-nucleon potential. This has, to some extent, been determined in a semi-phenomenological way and is thus somewhat ambiguous. It has been the hope of many nuclear physicists to use the results of nuclear matter calculations to discriminate between different potential models[43]. In the same spirit we may use the results of folded-diagram calculations on finite nuclei, such as ^{18}O, to assess the various potentials available. This is, however, not the main purpose of the present section. Here, our main concern is to discuss, once the nucleon-nucleon potential V is given, how to calculate nuclear spectra using the folded-diagram method.

From now on, we assume that the two-body part V of the nuclear Hamiltonian is given, for example by the Reid soft-core potential[39]. Then, as mentioned above, to calculate the energy spectrum of ^{18}O we need to calculate $v_{eff}(2)$ of eqs. (93) or (93.1). We shall use the latter. We first consider some specific classes of diagrams contained in the $\hat{Q}$-box, as shown in fig. 38.

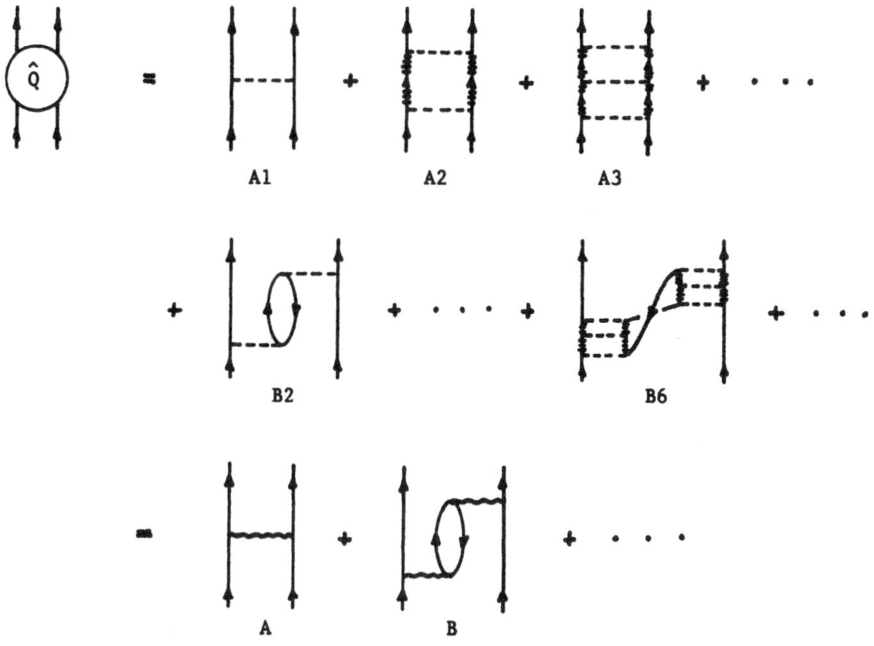

Fig. 38. $\hat{Q}$-box expressed in terms of G-matrix vertices, denoted by wavy lines ~~~~. Recall that the railed nucleon lines are the passive lines defined in fig. 1.

To handle the strong short-range repulsion contained in practically all nucleon-nucleon potential models, we need to use the Brueckner G-matrix[44]. Briefly speaking, this corresponds to a special scheme of partial summation of the $\hat{Q}$-box diagrams. Because of the strong repulsion at short distances, each vertex V of the diagrams shown in fig. 38 is very large (of the order of 1000 MeV, or infinite in the case of hard-core repulsion). Thus, each of the diagrams A1, A2, ... and B2, ... , B6, ... is very large. Consequently, we can not terminate the various sequences of diagrams at some finite order. For example, diagram A3 is of the same magnitude as diagram A2. We must therefore sum up the whole sequence $A1 + A2 + A3 + \cdots$. This can in fact be done, giving the G-matrix $\hat{Q}$-box diagram A shown in the last line of fig. 38. As the details of this partial summation procedure have been given elsewhere[44], we shall only point out that the G-matrix vertices are related to those of V by the integral equation

$$G = V + VQ_2 \frac{1}{\omega - Q_2TQ_2} Q_2G , \tag{94}$$

where Q_2 is the projector for the passive two-nucleon subspace, ω is the energy variable and T is the two-nucleon kinetic energy operator. It is clearly seen that G contains V vertices to all orders, with the restriction that the intermediate two-nucleon states must reside within Q_2 . Thus, the sequence A1, A2, A3, ... of fig. 38 is summed to all orders, giving rise to diagram A. Similarly, the sequence B1, ... , B6, ... is summed to all orders, giving diagram B. It may be mentioned that the matrix elements $\langle n_1n_2|G|n_3n_4\rangle$, the n_i's being harmonic oscillator wave functions, are typically of the magnitude of about 5 MeV. Thus, the G-matrix partial summation has served the purpose of regularizing the singular behaviour of V due to its strong repulsive core.

In fig. 39 we show several low-order $\hat{Q}$-box diagrams expressed in terms of G-matrix vertices. Shurpin et al.[45] have calculated $v_{eff}(2)$ of eq. (93.1)

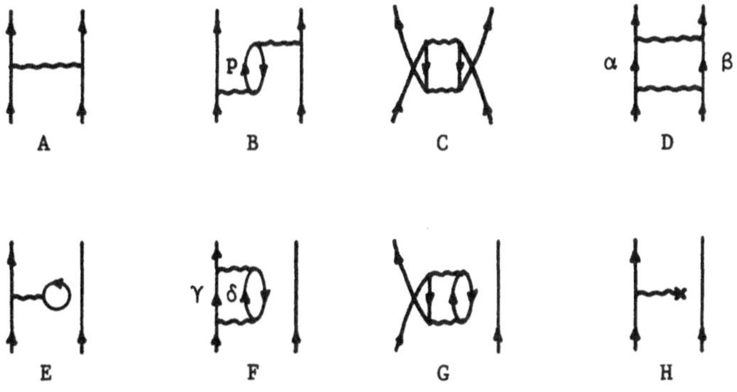

Fig. 39. Low-order $\hat{Q}$-box diagrams expressed in terms of antisymmetrized G-matrix vertices. Note that the intermediate states α, β, γ and δ of diagrams D and F must be chosen in consistence with the exclusion operator Q_2 [see eq. (94)] used in the G-matrix calculation[44]. It is understood that for a given diagram, all topologically distinct contributions are included. Thus, diagrams E - H contain contributions in which the insertions are made in the left single-particle line, as shown, and in the right single-particle line. Similarly, diagram B corresponds to the four contributions shown in fig. 35.

Table 2. Diagonal $\hat{Q}$-box diagrams of fig. 39. Both external lines have $j = 0d_{5/2}$ and are coupled to $J = 0$ and $T = 1$. The diagrams were calculated by Shurpin et al.[45] using the Reid nucleon-nucleon potential. Results are given for different values of the starting energy ω assigned to the pair of incoming lines. All entries are in units of MeV.

Diagram	Starting energy ω				
	-6	-8	-10	-12	-14
A	-1.47	-1.46	-1.45	-1.43	-1.42
B	-0.94	-0.86	-0.78	-0.71	-0.66
C	-0.39	-0.36	-0.33	-0.31	-0.29
D	-0.29	-0.27	-0.25	-0.23	-0.21
E	-51.86	-51.54	-51.21	-50.90	-50.59
F	-6.72	-6.15	-5.66	-5.23	-4.86
G	2.47	2.27	2.11	1.96	1.83

using a $\hat{Q}$-box given by the diagrams shown in fig. 39. Note that the $\sim\!\sim\!\times$ vertex of diagram H is the $-U$ vertex, because an unperturbed Hamiltonian $H_o = T + U$ with an auxiliary potential U is used, and consequently the interaction Hamiltonian becomes $H_1 = V - U$. Some typical values of these diagrams are shown in table 2. In these calculations an auxiliary potential $U = \frac{1}{2}m\omega^2r^2 + \Delta$ was used, with $\Delta = -54$ MeV and $\hbar\omega = 14$ MeV. This value of Δ was chosen because it gives a single-particle energy of -5 MeV for the (1s0d) shell, close to the empirical neutron separation energy of ^{17}O.

The calculations of Shurpin et al.[45)] were based on the folded-diagram method described in the present paper. They differ from the previous realistic nuclear structure calculations, such as those of Kuo and Brown[13)], Barrett and Kirson[15)], Grillot and McManus[46)], Barrett, Hewitt and McCarthy[47)] and Vary and Yang[48)], mainly in the following two aspects:

(i) Treatment of the folded diagrams.

(ii) Treatment of the Pauli exclusion operator in the calculation of the G-matrix.

Let us discuss point (ii) first. Shurpin et al. used the G-matrix elements calculated by Krenciglowa et al.[44)] where the G-matrix equation (94) was solved essentially exactly. Although the numerical values of the G-matrix elements so calculated differ from the earlier matrix elements only by about 5-15%, the present set of matrix elements is conceptually much more satisfactory. This is because we no longer have to worry about the so-called double-counting problem. If the Pauli exclusion operator Q_2 of eq. (94) is not treated accurately, there may be double counting between the nucleon-nucleon correlations already included in the G-matrix vertices and those to be included through matrix diagonalization within the model space P. The calculations of Krenciglowa et al. were based on a method suggested earlier by Tsai and Kuo[49)] and which allows an accurate and convenient treatment of the Q_2 projector by way of matrix inversion techniques. This method is convenient for light nuclei such as those in the (1s0d) shell. The method is still applicable to nuclei in the (1p0f) shell, and such a calculation has recently been carried out by Sommermann[50]. For heavier nuclei, however, the calculations become rather cumbersome.

Let us then discuss point (i) above. Typical contributions of folded diagrams to $v_{eff}(2)$ are shown in table 3. Here, F_n is the contribution from diagrams with n folds, as explained in eq. (93.1). Two observations can be made from table 3. Firstly, the folded diagrams are in fact important and cannot be ignored. Secondly, the folded-diagram series seems to be rapidly convergent when ordered according to the number of folds. Thus, it is sufficient to include diagrams with up to a few folds only. In previous calculations folded diagrams were either neglected[13] or were evaluated with the omission of some important terms[15,48,51]. As shown in table 3, the largest folded-diagram corrections come from F_1. Among the various diagrams contained in F_1 the two diagrams shown in fig. 40 are the most important ones. These are obtained by folding diagram A of fig. 39 with diagrams E and F, respectively. The diagram shown in fig. 40a deserves further comment.

Table 3. Contributions from folded diagrams, as given by Shurpin et al.[45]. Tabulated are the matrix elements $\langle aaJT|v_{eff}(2)|bbJT\rangle$ obtained from diagrams with different numbers of folds. The orbital labels 4, 5 and 6 denote $0d_{5/2}$, $0d_{3/2}$ and $1s_{1/2}$, respectively. The last column contains the matrix elements of Chung and Wildenthal[52]. All matrix elements are given in units of MeV.

T	J	a	b	F_0	F_1	F_2	F_3	F_4	Sum	CW
1	0	4	4	−2.801	0.969	−0.311	0.047	0.007	−2.089	−2.01
		4	5	−4.128	0.961	0.071	−0.009	−0.031	−3.136	−3.89
		4	6	−1.190	0.361	−0.070	−0.005	0.007	−0.897	−1.32
		5	5	−0.853	−0.063	0.028	0.044	0.013	−0.832	−0.81
		5	6	−0.912	0.281	0.008	−0.017	−0.007	−0.646	−0.84
		6	6	−2.131	0.810	−0.083	−0.061	0.019	−1.446	−2.31
		5	4	−4.128	1.380	−0.008	−0.051	−0.041	−2.849	
		6	4	−1.190	0.477	−0.138	0.014	0.005	−0.832	
		6	5	−0.912	0.216	0.013	0.002	0.008	−0.689	

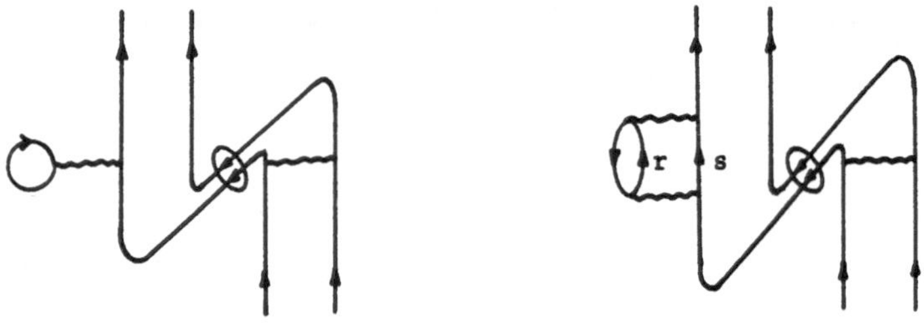

(a) (b)

Fig. 40. Two important once-folded diagrams.

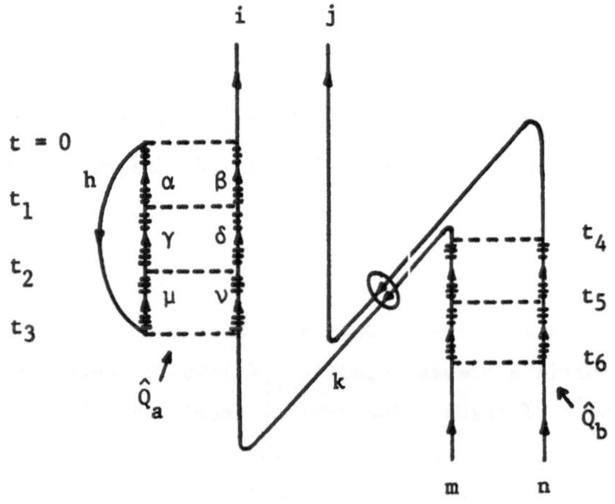

Fig. 41. A constituent diagram of diagram (a) of fig. 40.

According to eq. (49) once-folded diagrams come from the term $\hat{Q}'\int\hat{Q}$. Since $\hat{Q}'$ is at least second order in V , it follows that folded diagrams are at least third order in V . This rule is obviously not violated by the "second order" diagram of fig. 40a, since this diagram is written in terms of G-matrix vertices, and these are themselves composed of V vertices to all orders. In fact, the diagram of fig. 40a is composed of diagrams of the type shown in fig. 41. Note that the intermediate states α , β , γ , ... belong to the passive two-nucleon subspace defined by the projector Q_2 of the G-matrix equation (94), whereas the intermediate states r and s of diagram (b) in fig. 40 belong to the active two-nucleon subspace $P_2 \equiv 1 - Q_2$. Hence, there is no double counting between diagrams (a) and (b) of fig. 40, and both should be included.

The evaluation of diagram (a) of fig. 40 can be conveniently done using the energy derivative method discussed in sect. 5.4. Consider again the constituent diagram shown in fig. 41. The sum of all the diagrams of this type obeying the time constraints

$$0 > t_1 > t_2 > t_3 > -\infty ,$$
$$0 > t_4 > t_5 > t_6 > -\infty ,$$
$$0 > t_4 > t_3 , \qquad (95)$$

is readily found to be $-\frac{d\hat{Q}_a}{d\omega}\hat{Q}_b$, evaluated at $\omega = W_P$, the degenerate model-space energy for the two valence particles. The $\hat{Q}$-boxes $\hat{Q}_a$ and $\hat{Q}_b$ are indicated in fig. 41. Note that $\hat{Q}_b$ starts with terms which are first power in V , while $\hat{Q}_a$ starts with terms which are second power in V . Thus, it appears that in $\hat{Q}_a$ we do not have enough terms to form the G-matrix series shown by eq. (94). We notice, however, that V is independent of ω and consequently $\frac{dV}{d\omega} = 0$. This enables us to sum up all the diagrams of the type shown in fig. 41, obtaining

$$\langle hi|\left[\frac{dG}{d\omega}\right]_{\omega_1}|hk\rangle\langle kj|G(\omega_2)|mn\rangle\ , \tag{96}$$

with $\omega_1 = \varepsilon_h + \varepsilon_k$ and $\omega_2 = \varepsilon_m + \varepsilon_n$. This is in fact the procedure used by Shurpin et al.[45] to calculate diagram (a) of fig. 40. In previous G-matrix calculations[15,48,51] of the valence effective interaction, this diagram was not included. However, Shurpin et al. found that diagrams (a) and (b) of fig. 40 are of about equal importance. The large contribution from folded diagrams shown in table 3 is in fact mainly coming from these two diagrams. Physically they represent the renormalization of the valence effective interaction by the one-body depletion factor[45]. Thus, the main effect of the folded diagrams is to renormalize the non-folded diagrams by a factor $(1 - 2\kappa)$, where κ is the one-body depletion factor which is usually about 0.15.

Let us briefly summarize the developments made so far. We have shown how to calculate the matrix elements $\langle abJ\Gamma|v_{eff}(2)|cdJT\rangle$ starting from a given nucleon-nucleon potential. This involves the calculation of the G-matrix elements and then the calculation of the $\hat{Q}$-box in terms of its diagrams. (A detailed description of the G-matrix calculation has been given by Krenciglowa et al.[44] and a convenient and general method for the evaluation of the $\hat{Q}$-box diagrams by Kuo et al.[53].) Having obtained the appropriate matrix elements of $v_{eff}(2)$, we can readily calculate the energy spectra and model-space wave functions of ^{18}O by diagonalizing the matrix of H_{eff} given by eq. (92). Note that this $v_{eff}(2)$ is also the effective two-body interaction for nuclei with more than two valence nucleons, such as ^{19}O and ^{20}Ne, as we have discussed in sect. 5.3. The effective Hamiltonians for these two nuclei are clearly $H_{eff} = \tilde{H}_o(V) + v_{eff}(2) + v_{eff}(3)$ and $H_{eff} = \tilde{H}_o(V) + v_{eff}(2) + v_{eff}(3) + v_{eff}(4)$, respectively.

The above theory is formally exact, and it provides a systematic and rather simple method for calculating the energies and model-space wave functions of nuclei starting from a given nucleon-nucleon potential V. The general structure of this theory is similar to that of the empirical shell model. It must be admitted, however, that in practice we can not calculate the effective interactions $v_{eff}(2)$, $v_{eff}(3)$, ... exactly. Thus, suitable approximations have to be made, as will now be discussed in some detail.

As mentioned above, the only approximations we need to make are in the calculation of the $\hat{Q}$-box. For a given $\hat{Q}$-box, the folded-diagram series for v_{eff} can be summed to all orders using either the partial summation method of refs.[18,54] or the iteration method of Lee and Suzuki[55]. An elementary discussion of these two methods is given in ref.[56]. The folded-diagram series may also be ordered and calculated according to the number of folds, as indicated in table 3. The results of table 3 show rapid convergence in terms of the number of folds. However, such a rapid convergence cannot be guaranteed in this method. In general, both the partial summation method and the iteration method seem to have better convergence properties than a summation according to the number of folds[56]. Hence, it may be said that for a given $\hat{Q}$-box there should be no difficulty in calculating the folded $\hat{Q}$-box series.

Let us then discuss the approximations commonly made in calculating the $\hat{Q}$-box itself. These are of two types:

(i) Neglect of higher order $\hat{Q}$-box diagrams.

(ii) Neglect of contributions from certain high-energy intermediate states in individual $\hat{Q}$-box diagrams.

We use fig. 42 to explain the meaning of these approximations. Here, diagram B is a $\hat{Q}$-box diagram which is second order in G, as shown in fig. 38. It is the well known lowest order core-polarization diagram. In several early calculations[13,14,57] of the core-polarization contributions to v_{eff}, only diagram B was included. There are, however, many other $\hat{Q}$-box diagrams which are of core-polarization nature. Examples are diagram

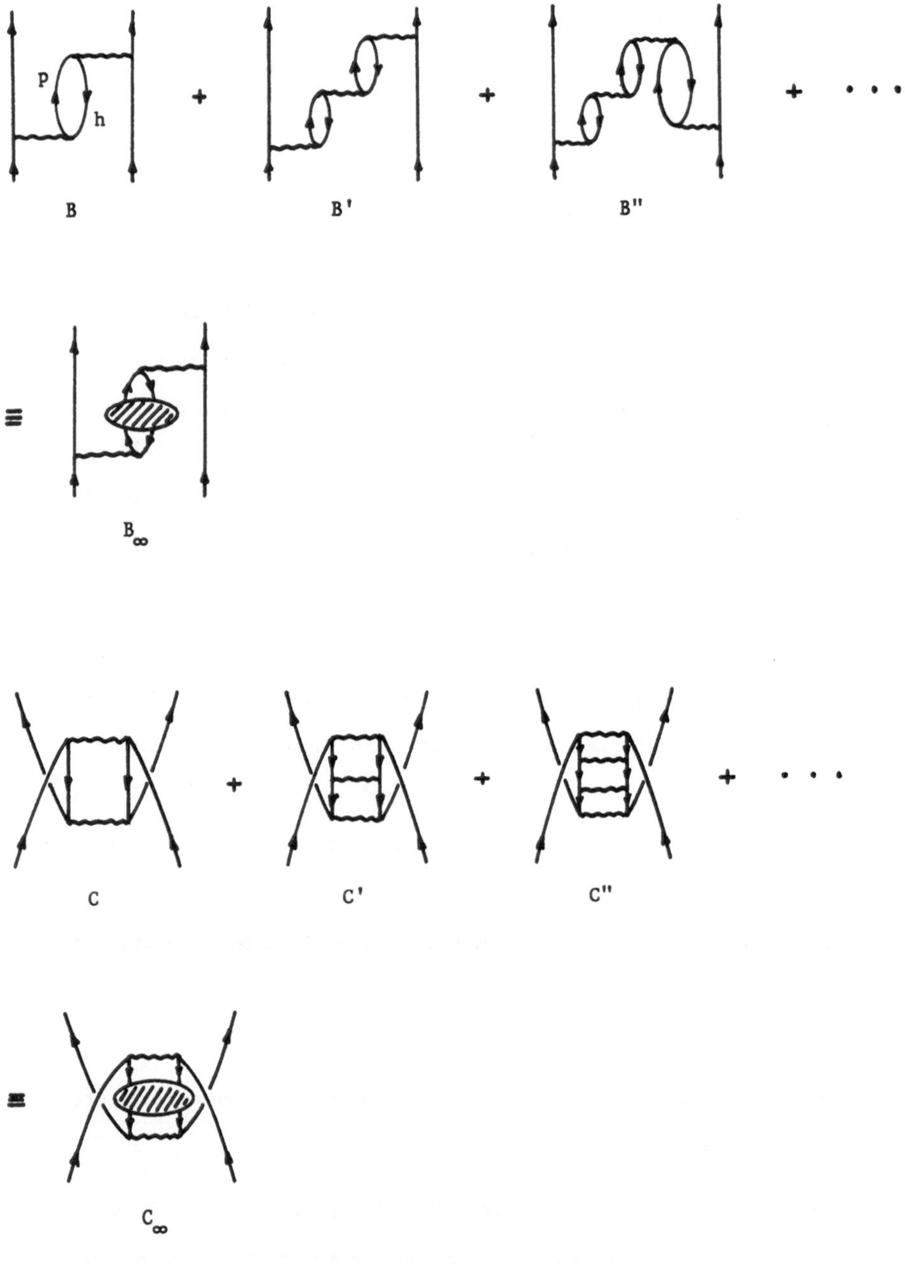

Fig. 42. Some higher order $\hat{Q}$-box diagrams.

B' (third order in G) and diagram B" (fourth order in G). How important are these diagrams? This question has been investigated by many authors [see e.g. the review articles of Barrett and Kirson[10], Kuo[11], Ellis and Osnes[12] and Andō, Bandō and Nagata[58]]. Unfortunately, the answer to this question is not yet conclusive. As expected, there is no general evidence that the $\hat{Q}$-box diagrammatic series is rapidly convergent when grouped according to the power of the G-matrix vertices. The subset of diagrams B, B', B", ... can easily be summed to all orders, and the result B_∞ is in general quite different in magnitude from diagram B [12]. Similar observations[59] are made for the subset of diagrams C, C', C", ... Here, diagram C is the lowest order hole-hole correlation diagram, while in diagrams C' and C" repeated hole-hole correlations are allowed. Thus, the question arises, whether we should sum the $\hat{Q}$-box series order by order in G or calculate certain classes of $\hat{Q}$-box diagrams to all orders by way of partial summation.

Frankly speaking, we do not know the answer to this question. However, recent folded-diagram calculations on A = 18 - 20 nuclei by Shurpin et al.[59] have thrown new light on this problem. They used the Reid[39] and Paris[41] nucleon-nucleon potentials and found, for both potentials, that the spectra obtained with diagrams B and C included in v_{eff} were quite different from those obtained with diagrams B and C replaced by diagrams B_∞ and C_∞, when folded diagrams were not included. However, when folded diagrams were included, as they should, the results of these two cases were in fact quite similar. These findings seem to indicate that the inclusion of appropriate folded diagrams serves to suppress the importance of higher order $\hat{Q}$-box diagrams. Further studies will be necessary, however, in order to test the general validity of these interesting results.

We then discuss the approximations of type (ii) , dealing with the summation over intermediate states in the $\hat{Q}$-box diagrams. Again, we use diagram B of fig. 42 as an example. In this diagram, the intermediate-state indices to be summed over are p and h . The summation over h is finite, as there are only three orbits for h in the ^{16}O core, namely $0s_{1/2}$, $0p_{3/2}$ and $0p_{1/2}$. But the summation over p is infinite, including all the single-particle orbits from $0d_{5/2}$ and on. This infinite summation introduces serious practical complications to the evaluation of this diagram. In the early calculations[13,57)], this summation was truncated at $2\hbar\omega$ harmonic oscillator energy, i.e. the infinite summation was replaced by a finite summation enforced by the restriction of $\varepsilon_p - \varepsilon_h \leq 2\hbar\omega$, where ε_p and ε_h are the single-particle energies of the orbits p and h . This approximation simplified the calculation considerably, and furthermore, it seemed intuitively reasonable. In fact, the results so obtained were quite promising[13,57)].

Consequently, the question arises, if the above truncation of the intermediate-state summation is justified. A critical study of this question was carried out by Vary, Sauer and Wong (VSW)[60)]. They arrived at the disturbing conclusion that this truncation is not justified. Denoting by B(n) the contribution to diagram B of fig. 42 from particle-hole excitations with $\varepsilon_p - \varepsilon_h = n\hbar\omega$, they found that B(n) was quite important for $n > 2$. In fact, the intermediate-state summation converged only when particle-hole excitations up to $22\hbar\omega$ were included. Furthermore, there was a strong cancellation between B(2) and $\sum_{n>2}^{22} B(n)$, so that the net result was $B \approx \sum_{n=2}^{22} B(n) \approx 0$. This result is indeed very embarrassing, for two reasons. Firstly, the calculation of $\hat{Q}$-box diagrams, of which diagram B is one, would become exceedingly complicated, because we should have to sum over intermediate particle states to very high energies. Secondly, the final result of diagram B itself is very small and thus unable to provide the long-range quadrupole interaction obtained if only $2\hbar\omega$ particle-hole excitations are included[13,57)].

Recently, Kung, Kuo and Ratcliff[61)] performed an independent study of the above VSW effect. They used a plane-wave representation for the high-lying intermediate particle states p of diagram B and a harmonic oscillator representation for the low-lying p states. The plane-wave states were properly orthogonalized to the oscillator states to avoid double counting. The use of a plane-wave representation rather than a harmonic oscillator representation for the high-lying intermediate particle states was motivated by the following considerations. In the oscillator representation the intermediate-state summation is discrete and must be truncated at some finite energy, e.g. $\varepsilon_p - \varepsilon_h = 22\hbar\omega$. In the plane-wave representation the summation over intermediate particle energies is replaced by integration, for each partial wave, over the particle momentum. Using a suitable set of momentum-space mesh points, this integration can be performed essentially exactly over the entire momentum range of p. Another reason for using plane waves for p was to study the effect of using different single-particle potentials for the high-lying intermediate states. The final results of Kung et al. were, however, both qualitatively and quantitatively similar to those obtained by VSW.

The above slow convergence of the intermediate-state summation is usually referred to as the VSW effect. This effect has for a long time been of major concern to studies of the convergence properties of realistic nuclear effective interactions, as pointed out above. A recent calculation by Sommermann et al.[62)] points, however, towards a different conclusion. As noted in refs.[60,61)], the VSW effect is closely related to the strong tensor-force component in the free nucleon-nucleon potential. Thus, a weaker tensor force might well reduce the VSW. This is precisely what Sommermann et al. have found. The tensor force in the nucleon-nucleon potential comes mainly from the exchange of a single pion (π) and a single rho-meson (ρ). It is worth noting that these two contributions are of opposite signs and will partly cancel each other, depending on the relative strengths of the π-nucleon and ρ-nucleon coupling constants. Sommermann et al. used a recent Bonn-Jülich potential[40)] which has a weak tensor force because a strong ρ-nucleon coupling constant was employed, as suggested by

Höhler and Pietarinen[63]. In fact, this potential reproduces the two-nucleon data equally well as the other modern models of the nucleon-nucleon potential. Surprisingly, and to the satisfaction of many, Sommermann et al. found that the core-polarization diagram B of fig. 42 can be calculated to good accuracy by restricting the intermediate-state summation to particle and hole states with $\varepsilon_p - \varepsilon_h = 2\hbar\omega$. Furthermore, they showed that this truncation was associated with the weak tensor force of the Bonn-Jülich potential. It would thus be interesting, as well as useful, to see their investigation extended to other $\hat{Q}$-box diagrams and to other nucleon-nucleon potentials such as that of the Paris group[41].

In summary, the folded-diagram theory provides a convenient and formally exact method for calculating properties of nuclei like ^{18}O and ^{19}O, starting from the free nucleon-nucleon interaction. The effect of the folded diagrams themselves is quite important. However, approximations have to be made in the evaluation of the $\hat{Q}$-box. These consist of leaving out both complicated higher order $\hat{Q}$-box diagrams which cannot easily be calculated and the contributions from high-energy intermediate states. It is probable that the use of a nucleon-nucleon potential with a weak tensor force and the inclusion of folded diagrams will improve the accuracy of such calculations.

During recent years, several folded-diagram calculations of nuclear structure properties, using approximate $\hat{Q}$-boxes, have been performed[45,59,62,64]. Typical examples of nuclear energy spectra obtained in such calculations are shown in figs. 43-45. In fig. 43 are shown the ^{18}O spectra obtained by Sommermann et al.[62] for two different potential models of the nucleon-nucleon interaction. Similar spectra of ^{17}O were calculated by Tam et al.[64] and are shown in fig. 44, while the effects on the ^{17}O spectra of using different single-particle spectra and $\hat{Q}$-boxes are studied in fig. 45. Although the agreement between the calculated and experimental spectra is not perfect, the results are promising enough to warrant further, more complete folded-diagram calculations.

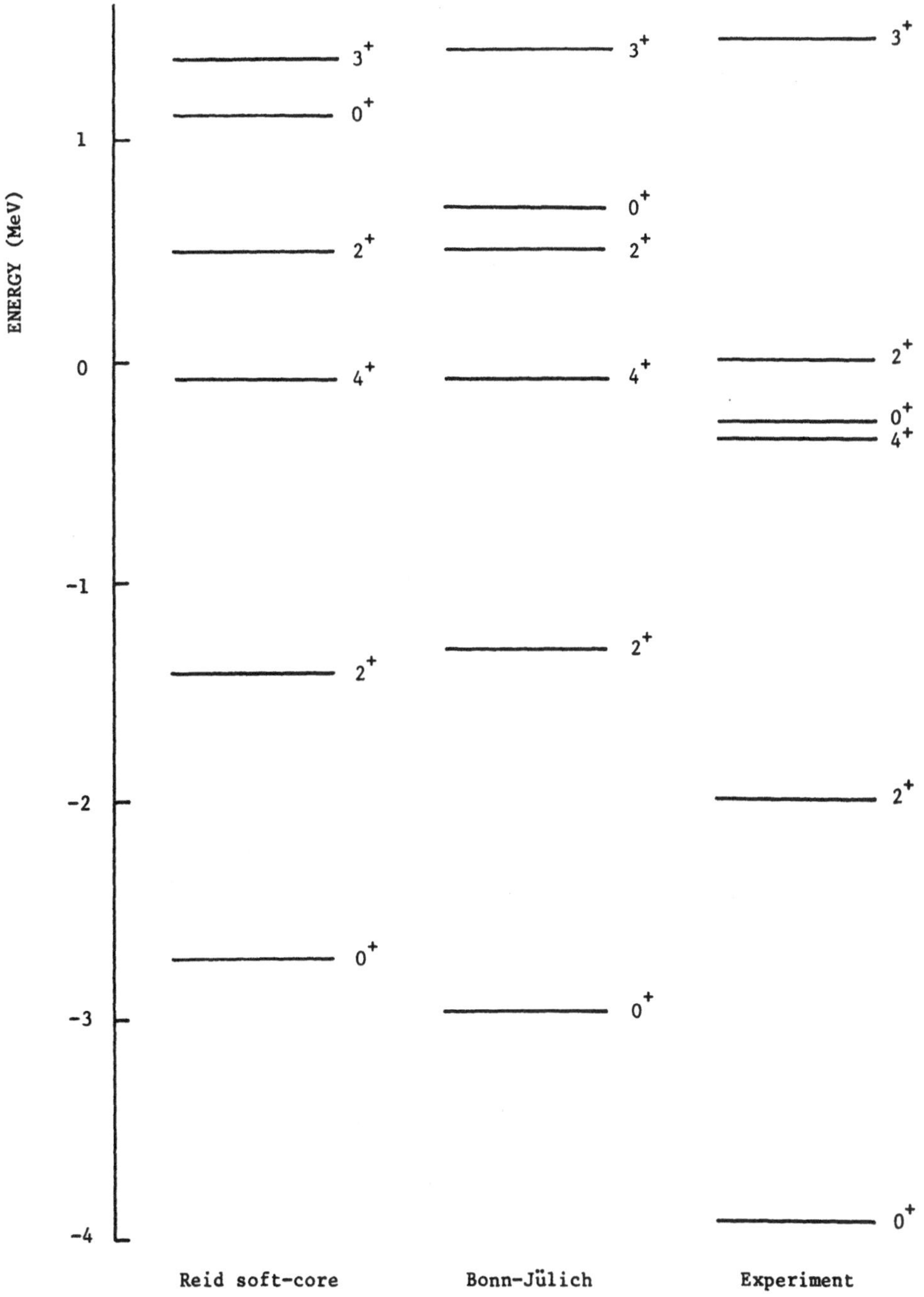

Fig. 43. Energy levels of ^{18}O, as calculated by Sommermann et al.[62]. A $\hat{Q}$-box composed of diagrams A-H of fig. 39 was used in the folded-diagram expansion of $v_{eff}(2)$. High-energy intermediate p states were included in diagram B. Note that the results given by the two potentials are quite similar, with the Bonn-Jülich potential providing slightly more binding.

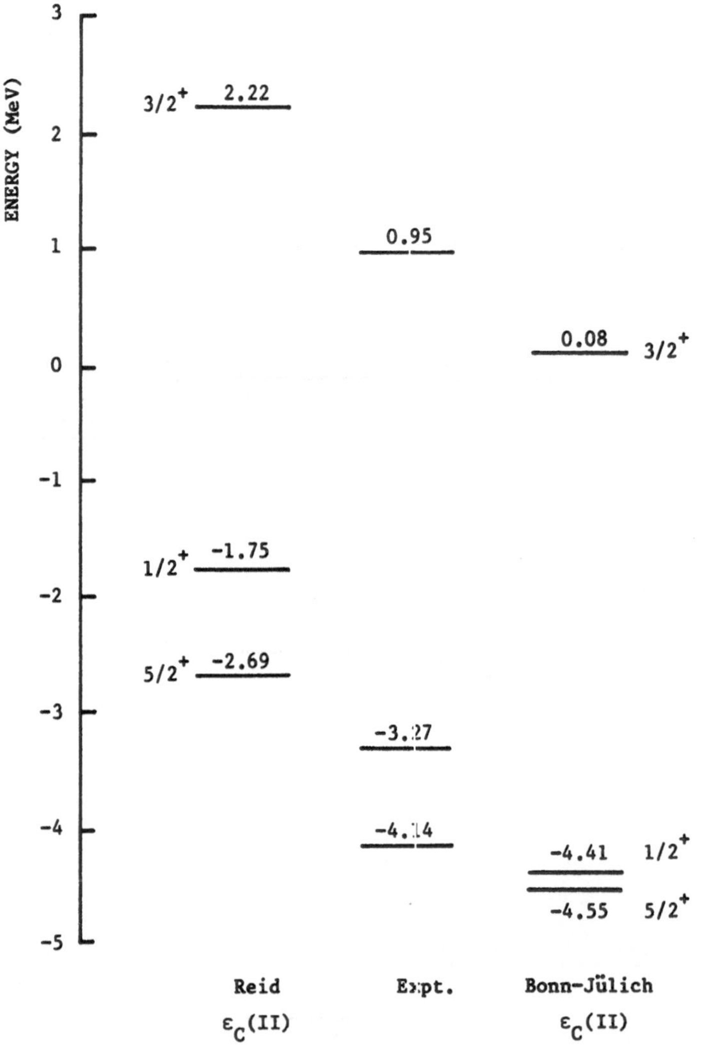

Fig. 44. Folded-diagram calculations[64)] of the ^{17}O spectrum for two different nucleon-nucleon potentials. See the caption of fig. 45 for further explanations.

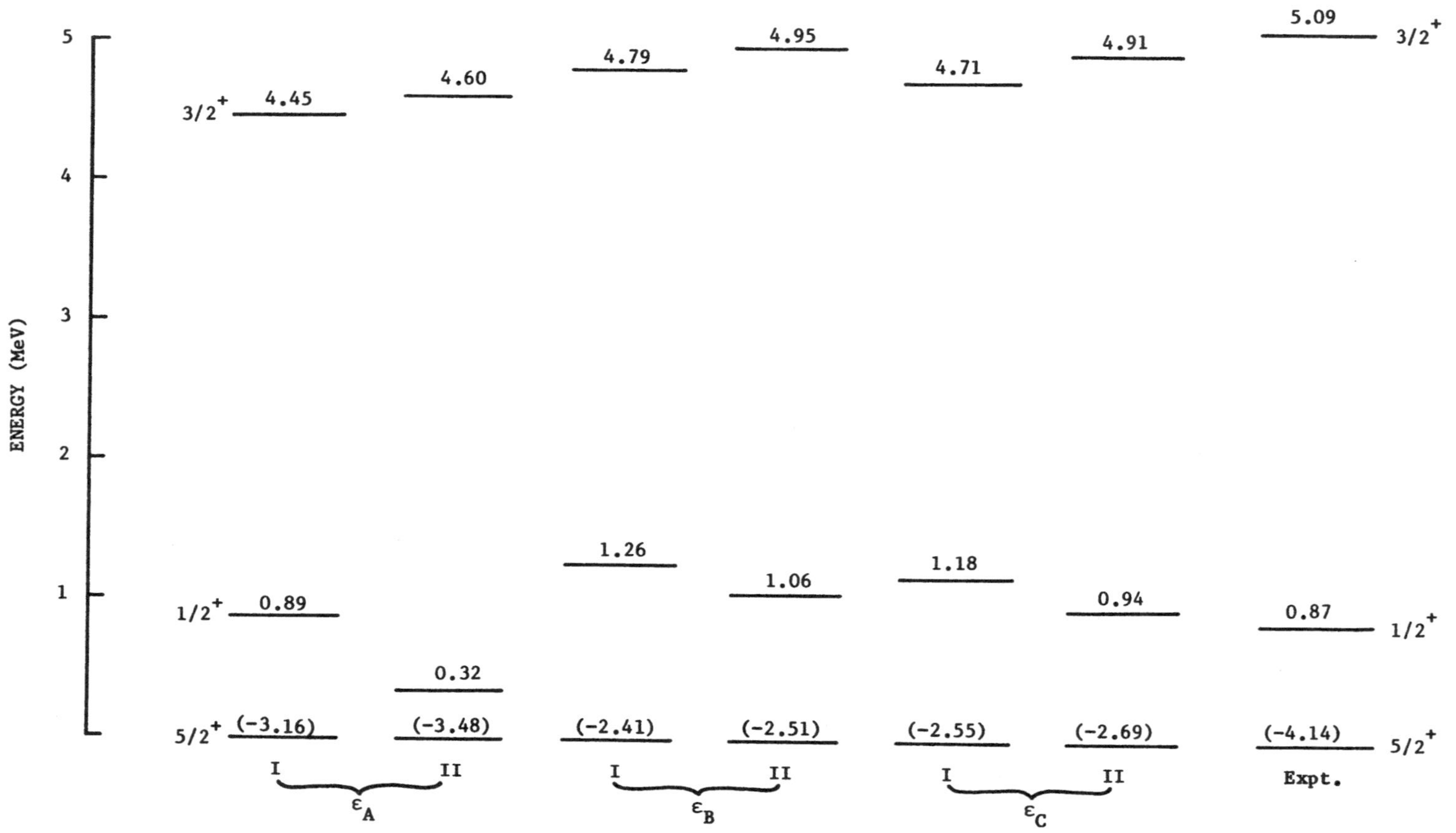

Fig. 45. Folded-diagram calculations[64)] of the ^{17}O spectrum for two different $\hat{Q}$-boxes (I and II) and three different single-particle spectra (ε_A , ε_B and ε_C) . The $\hat{Q}$-box (I) is of second order, while (II) contains particle-hole phonons. Folded diagrams were included to all orders for these $\hat{Q}$-boxes. Note the similarity of the calculated spectra.

8. Further nuclear applications

The basic purpose of the folded-diagram theory developed in the previous sections can be stated rather concisely. Given a many-body Hamiltonian H and a chosen model space P, this theory describes a systematic method to derive an effective Hamiltonian H_{eff}, which is defined within the P-space (i.e. $H_{eff} = PH_{eff}P$) and reproduces certain eigenvalues of H and the P-space projections of the corresponding eigenfunctions. Since P is usually a very restricted space, the degrees of freedom allowed by H_{eff} are usually much fewer in number than those allowed by the original Hamiltonian H. Thus, calculations with the Schrödinger equation defined by H_{eff} are generally much simpler, and moreover, the physical meaning of its solutions is usually much more conceivable and desirable than those of the original Schrödinger equation defined by H.

The folded-diagram derivation of H_{eff} described in the previous sections is in fact very general. So far we have applied it only to the derivation of the nuclear shell-model effective Hamiltonian. It is, however, clear that the folded-diagram theory can be applied to other physical problems as well. The recent works[30,31] on the folded-diagram derivation of the nuclear optical model potential and of the free nucleon-nucleon potential seem to be both interesting and promising. These developments will be discussed briefly in the following subsections.

8.1. Nuclear optical model potential.

Let us consider the elastic scattering of one nucleon by a nucleus consisting of A nucleons. The ground state of this nucleus is denoted by Ψ_o^A and its energy by E_o^A. The Schrödinger equation of the total scattering system of $A+1$ nucleons is

$$H \Psi^{A+1} = E^{A+1} \Psi^{A+1} , \tag{97}$$

which is, as we know, very difficult to solve. The elastic scattering of one nucleon by the nucleus A corresponds to a special choice of the model space, namely with basis vectors of the form

$$|\Phi_{\vec{k}}\rangle \equiv a^{\dagger}_{\vec{k}} |\Psi^A_o\rangle , \tag{98}$$

obtained by adding a nucleon in a single-particle state with momentum $\vec{k}$ to the true ground state of the target nucleus A. The projector for this model space is

$$\bar{P} = \int d\vec{k} \, |\Phi_{\vec{k}}\rangle\langle\tilde{\Phi}_{\vec{k}}| , \tag{98.1}$$

where, because of the presence of the true ground state Ψ^A_o in eq. (98), the wave functions $\Phi_{\vec{k}}$ fulfil the biorthogonality relation

$$\langle\tilde{\Phi}_{\vec{k}} | \Phi_{\vec{k}'}\rangle = \delta(\vec{k} - \vec{k}') . \tag{98.2}$$

Thus, for the elastic scattering problem, we are interested only in the projections of the wave functions Ψ^{A+1} onto the model space $\bar{P}$. In other words, we look for the model-space Schrödinger equation

$$\bar{H}_{eff}\bar{P}\Psi^{A+1} = E^{A+1}\bar{P}\Psi^{A+1} , \tag{99}$$

where $\bar{H}_{eff}$ is operating only within the model space $\bar{P}$, i.e. $\bar{H}_{eff} = \bar{P}\bar{H}_{eff}\bar{P}$.

The transition from the original Schrödinger equation (97) to the model-space equation (99) is very similar to the derivation of the effective Hamiltonian discussed in detail in sect. 4. There is, however, one major difference. In order to derive the shell-model effective Hamiltonian of, say, ^{17}O using the method described in sect. 4, one would employ a model space whose basis vectors were of the form

$$|\Phi'_{\vec{k}}> \;=\; a^{\dagger}_{\vec{k}}\,|\Phi^A_o> \;, \tag{100}$$

where Φ^A_o is the unperturbed ground state of the core nucleus ^{16}O. The corresponding projection operator is then

$$P' \;=\; \int d\vec{k}\; |\Phi'_{\vec{k}}><\Phi'_{\vec{k}}| \;. \tag{100.1}$$

The difference between $\Phi_{\vec{k}}$ and $\Phi'_{\vec{k}}$ (or $\overline{P}$ and P') is significant, because Ψ^A_o is the true ground state of the nucleus A, in the present case ^{16}O, while Φ^A_o is the corresponding unperturbed ground state. The state Φ^A_o is chosen by us, and hence its structure is known in detail. On the other hand, the detailed structure of Ψ^A_o is in general unknown. Then, how do we proceed to derive $\overline{H}_{eff}$ from H and the unknown $\overline{P}$?

Here, we use an equation of motion method which was developed by Kuo and Krenciglowa[65)†] [see also the discussion of ref.[56)]] to deal with this type of problem. This method was subsequently applied to the derivation of $\overline{H}_{eff}$ for the elastic scattering problem by Kuo, Osterfeld and Lee[30)]. They started from the equation of motion

$$<\Psi^A_o|[a_{\vec{k}},H]|\Psi^{A+1}_E> \;=\; (E^{A+1} - E^A_o)<\Psi^A_o|a_{\vec{k}}|\Psi^{A+1}_E> \;, \tag{101}$$

where Ψ^{A+1}_E is the exact wave function of the $A+1$ system with the appropriate scattering boundary condition. In elastic scattering, as mentioned earlier, we observe only a specific part of the exact wave function Ψ^{A+1}_E, namely the part with the target in its ground state Ψ^A_o and the projectile in a plane wave state $\vec{k}$. This is just the usual optical model wave function $\rho_E(\vec{k})$ given by

$$\rho_E(\vec{k}) \;\equiv\; <\Psi^A_o|a_{\vec{k}}|\Psi^{A+1}_E> \;. \tag{101.1}$$

---

† In this reference some connections between the folded-diagram theory and the Green's function many-body theory are discussed.

Our aim is to determine an effective Hamiltonian from which $\rho_E(\vec{k})$ can be determined. This can be accomplished by performing a folded-diagram analysis of the matrix element on the l.h.s. of eq. (101).

Using the complex-time approach employed in sect. 4, we can construct the bra vector of this matrix element by

$$<\Psi_o^A| \quad \propto \quad \frac{<\Phi_o^A|U(\infty,0)}{<\Phi_o^A|U(\infty,0)|\Phi_o^A>} \quad . \tag{101.2}$$

Similarly, we have for the ket vector

$$|\Psi_E^{A+1}> \quad \propto \quad \frac{U(0,-\infty)|\chi_E>}{<\chi_E|U(0,-\infty)|\chi_E>} \quad , \tag{101.3}$$

where χ_E is a parent state, similar to ρ_λ of eq. (28), constructed as a linear combinations of P'-states $a^{+}_{\vec{\mu}}|\Phi_o^A>$. Substituting these expressions into eq. (101), we have

$$\frac{1}{D}<\Phi_o^A|U(\infty,0)A_{\vec{k}}U(0,-\infty)|\chi_E> \quad = \quad \frac{1}{D}(E^{A+1} - E_o^A - \varepsilon_{\vec{k}}) <\Phi_o^A|U(\infty,0)a_{\vec{k}}U(0,-\infty)|\chi_E> \ , \tag{101.4}$$

where we have used the notations

$$D \quad \equiv \quad <\Phi_o^A|U(\infty,0)|\Phi_o^A> <\chi_E|U(0,-\infty)|\chi_E> \tag{101.5}$$

and

$$A_{\vec{k}} \quad \equiv \quad [a_{\vec{k}}, H] \ - \ \varepsilon_{\vec{k}}a_{\vec{k}} \ . \tag{101.6}$$

In the last equation the term $\varepsilon_{\vec{k}}a_{\vec{k}}$ is the commutator $[a_{\vec{k}}, H_o]$ and thus $A_{\vec{k}}$ is just the commutator $[a_{\vec{k}}, H_1]$, since $H = H_o + H_1$.

Now, the unlinked diagrams, i.e. those containing pieces which are not linked to the valence line, can be factorized out from eq. (101.4), giving

$$\frac{N_o}{D} \langle \Phi_o^A | U(\infty,0) A_{\vec{k}} U(0,-\infty) | \chi_E \rangle_L = \frac{N_o}{D} (E^{A+1} - E_o^A - \varepsilon_{\vec{k}}) \langle \Phi_o^A | U(\infty,0) a_{\vec{k}} U(0,-\infty) | \chi_E \rangle_L , \tag{101.7}$$

where

$$N_o \equiv \langle \Phi_o^A | U(\infty,-\infty) | \Phi_o^A \rangle \tag{101.8}$$

and the subscript L implies that only linked diagrams are included. Some typical diagrams belonging to $\langle \Phi_o^A | U A_{\vec{k}} U | \chi_E \rangle_L$ and $\langle \Phi_o^A | U a_{\vec{k}} U | \chi_E \rangle_L$ are shown in fig. 46. Note that for $\langle \Phi_o^A | U A_{\vec{k}} U | \chi_E \rangle_L$, all the diagrams have $A_{\vec{k}}$, denoted by a solid-line vertex, at the time $t = 0$, as shown by diagram (a). On the other hand, all the diagrams of $\langle \Phi_o^A | U a_{\vec{k}} U | \chi_E \rangle_L$ have the destruction operator $a_{\vec{k}}$ at the time $t = 0$. Note further that the vertices of H_1 may be anywhere between the times $t \to \infty$ and $t \to -\infty$. For example, we have $\infty > t' > t_1$ for both diagrams (a) and (b). As shown by the figure, the structure of these two diagrams indicates that we may factorize diagram (a) into "something" multiplied by diagram (b). This is precisely what we will do, and to carry out this factorization, we need to use the folded-diagram factorization procedure described in sect. 3.

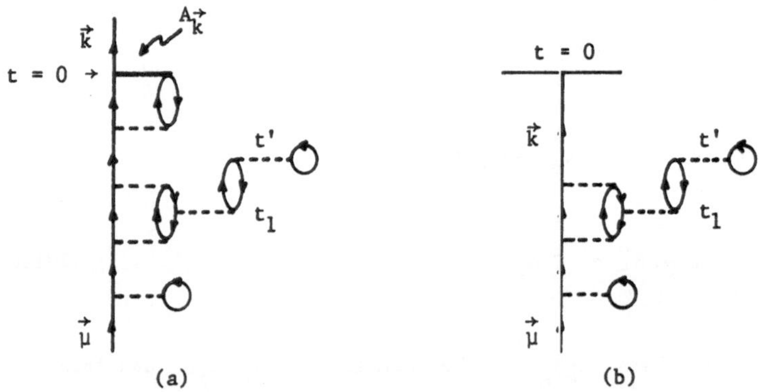

Fig. 46. Linked diagrams of eq. (101.7).

In fig. 47 we show the general diagrammatic structure of $<\Phi_o^A|Ua_{\vec{k}}U|\chi_E>_L$. Here, each circle represents a $\hat{Q}$-box, composed of irreducible diagrams. For example, diagram (b) of fig. (46) belongs to the term with two $\hat{Q}$-boxes. Similarly, the diagrammatic structure of $<\Phi_o^A|UA_{\vec{k}}U|\chi_E>_L$ is shown in fig. 48, where the slashed box indicates that every diagram of this box has one $A_{\vec{k}}$ vertex at the time $t = 0$. For example, diagram (a) of fig. 46 belongs to the term with one slashed $\hat{Q}$-box and two ordinary $\hat{Q}$-boxes.

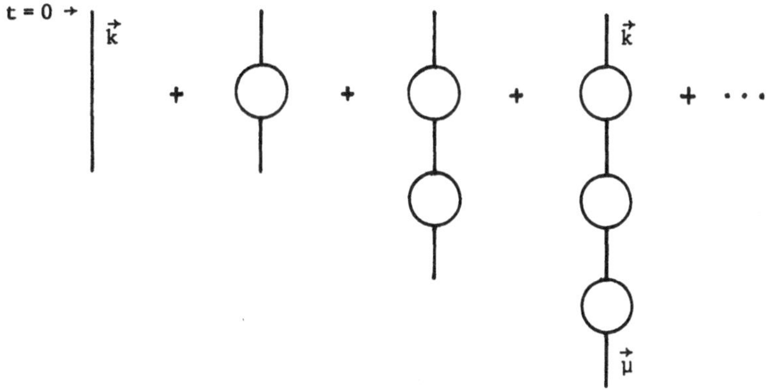

Fig. 47. Diagrammatic structure of $<\Phi_o^A|Ua_{\vec{k}}U|\chi_E>_L$.

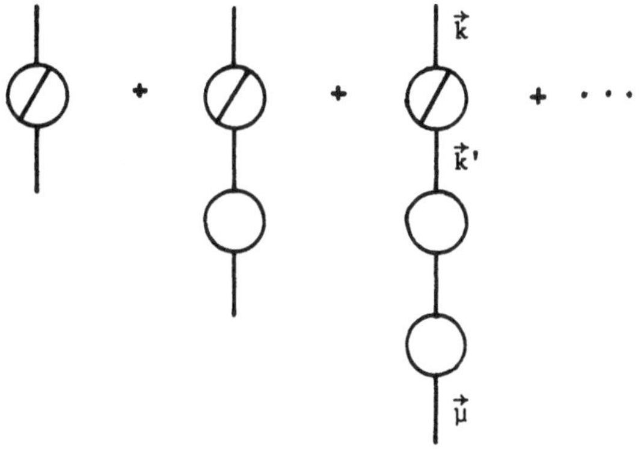

Fig. 48. Diagrammatic structure of $<\Phi_o^A|UA_{\vec{k}}U|\chi_E>_L$.

Let us now make a folded-diagram factorization of each term in $<\Phi_o^A|UA_{\vec{k}}U|\chi_E>_L$, as shown in fig. 49. Collecting terms columnwise, we readily have

$$<\Phi_o^A|UA_{\vec{k}}U|\chi_E>_L = \int d\vec{k}' \, U^{opt}_{\vec{k}\vec{k}'} <\Phi_o^A|Ua_{\vec{k}'}U|\chi_E>_L , \qquad (101.9)$$

where $U^{opt}_{\vec{k}\vec{k}'}$ is given by the linked diagram expansion shown in fig. 50. Then, substituting eq. (101.9) into eq. (101.7) and applying eqs. (101.3) and (101.1), we obtain

$$\varepsilon_{\vec{k}}\rho_E(\vec{k}) + \int d\vec{k}' \, U^{opt}_{\vec{k}\vec{k}'} \, \rho_E(\vec{k}') = (E^{A+1} - E_o^A) \, \rho_E(\vec{k}) , \qquad (102)$$

which is the non-local Schrödinger equation for the optical model wave function $\rho_E(\vec{k})$. Thus, $U^{opt}_{\vec{k}\vec{k}'}$ is the optical model potential we have been looking for.

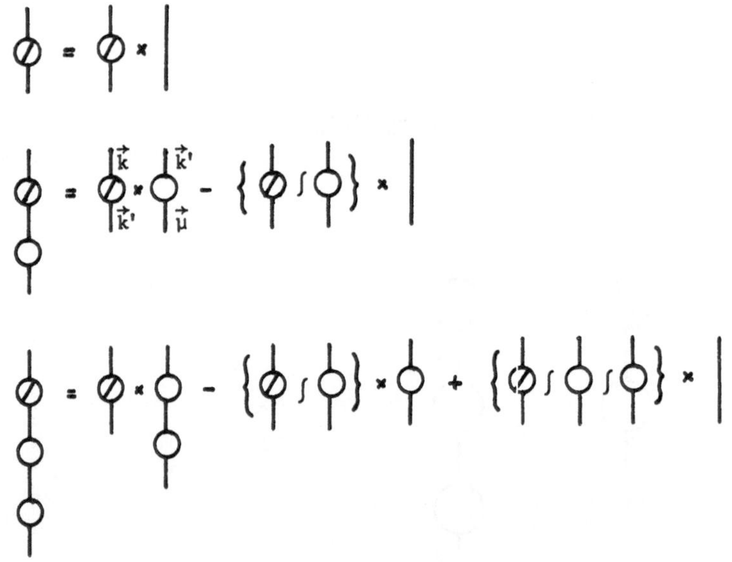

Fig. 49. Folded-diagram factorization of $<\Phi_o^A|UA_{\vec{k}}U|\chi_E>_L$.

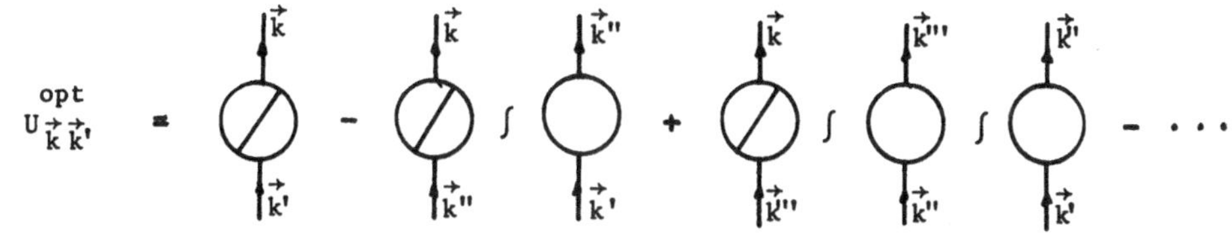

(a)

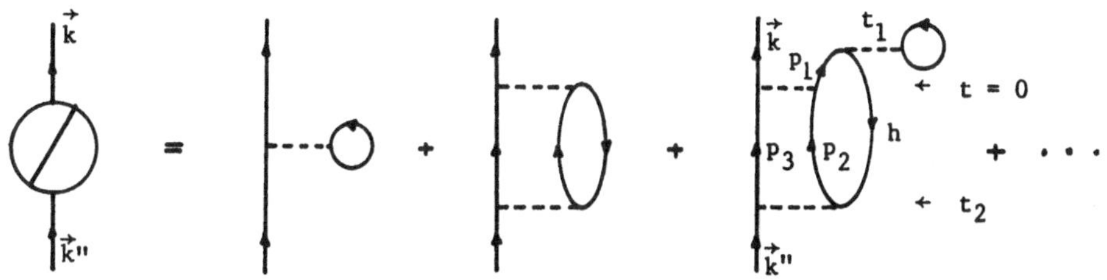

(b)

Fig. 50. Folded-diagram expansion of U^{opt}. Note that the vertex at the time $t = 0$ corresponds to the solid-line $A_{\vec{k}}$ vertex of fig. 46a.

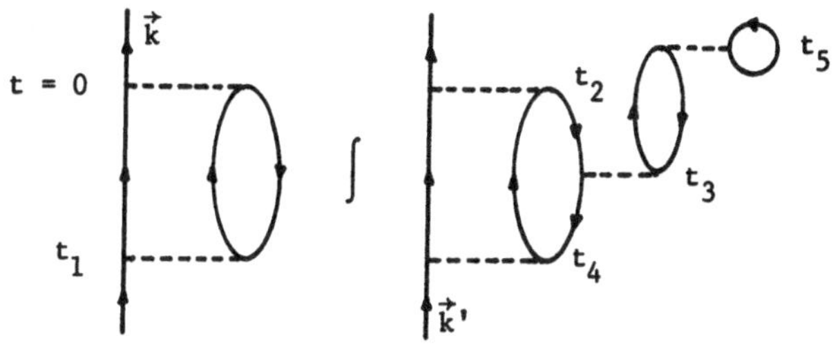

Fig. 51. A generalized folded diagram of U^{opt}.

Note that the diagrammatic expansion of the optical potential U^{opt} shown in fig. 50a is identical to the folded-diagram expansion of the effective interaction v_{eff} given by eq. (38a) and later by eq. (49), except for the presence of the slashed $\hat{Q}$-box. This is a nice feature of the present folded-diagram theory; it gives a unified formulation of the effective interaction for bound state and scattering problems. In fact, the slashed $\hat{Q}$-box is equal to the one-body $\hat{Q}$-box except for one subtle difference in the time integration. As shown in fig. 50b, the slashed $\hat{Q}$-box is composed of one-body irreducible and valence-linked diagrams. In the $\hat{Q}$-boxes of the previous sections, all the time variables were to be integrated over the time interval from $-\infty$ to 0. In the slashed $\hat{Q}$-box, however, the time variables are to be integrated from $-\infty$ to $+\infty$, noting that there is always a "fixed" vertex at $t = 0$ which is attached directly to the outgoing external valence line $\vec{k}$. Consider as an example the third diagram of fig. 50b. The time integrations of this diagram are $\int_0^\infty dt_1 \int_{-\infty}^0 dt_2$, leading to an energy factor $1/[(\varepsilon_h-\varepsilon_{p_1})(\varepsilon_{k''}-\varepsilon_{p_3}-\varepsilon_{p_2}+\varepsilon_h)]$. The reason for these peculiarities can be understood from eq. (101.4), where the two time evolution operators $U(\infty,0)$ and $U(0,-\infty)$ appear and the interaction vertex coming from H is fixed at the time $t = 0$.

It should be pointed out that the reference time for the folding operation in the folded-diagram expansion for U^{opt} is still $t = 0$. Let us explain this by way of the diagram drawn in fig. 51. This diagram belongs to the second term in the expansion of U^{opt} shown in fig. 50a. The reducible diagram, from which this folded diagram originates, has the same structure but a different time ordering, namely $0 > t_1 > t_2 > t_3 > t_4 > -\infty$ and $\infty > t_5 > t_3$. This reducible diagram is factorized into two parts, one with $0 > t_1 > -\infty$ and the other with $0 > t_2 > t_3 > t_4 > -\infty$ and $\infty > t_5 > t_3$. Thus, incorrect time ordering is introduced, and this is compensated for by the folded diagram. Note that the factorized diagram has $0 > t_2$, but <u>not</u> $\infty > t_2$. Hence, the folded diagram of fig. 51 has the generalized time ordering $0 > t_1 > -\infty$, $0 > t_2 > t_3 > t_4 > -\infty$, $\infty > t_5 > t_3$ and $t_2 > t_1$.

The physical meaning of the above theory may be visualized by fig. 52. The elastic scattering problem is represented by the processes shown in part (a) of the figure. The projectile-nucleon a is coming in towards the target nucleus, which is in its ground state Ψ_o^A. Then, complicated nuclear reactions take place. Eventually the projectile-nucleon a emerges, leaving the target in its ground state. Of course, many other processes may take place, such as $a + A \rightarrow a^* + A^*$ or $a + A \rightarrow b + B$, where a^* and A^* are excited states of a and A, and b and B are different nuclei from a and A. Thus, these other processes are considered to lead to $\overline{Q}$-space states. Elastic scattering measurements are conducted only inside the $\overline{P}$-space (recall that $\overline{P} + \overline{Q} = 1$) where the target is restricted to be in its ground state Ψ_o^A. In other words, we only measure a small fraction of the complicated wave function Ψ_E^{A+1}, namely the amplitude $\rho_E(\vec{k})$ defined by eq. (101.1). This amplitude is usually referred to as the optical model wave function. The present folded-diagram theory tells us that $\rho_E(\vec{k})$ can be calculated by solving the one-body Schrödinger equation (102), where U^{opt} is an effective one-body potential given by fig. 50. This is an interesting and potentially important result.

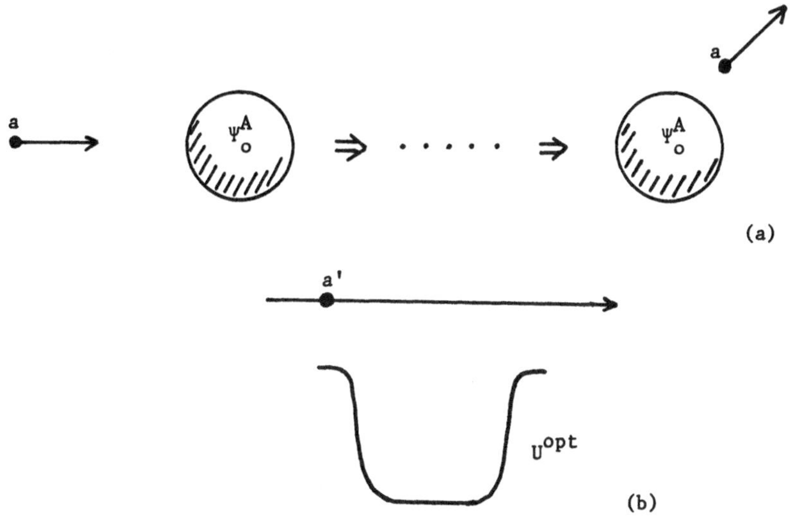

Fig. 52. Schematic picture of elastic $a + A \rightarrow a + A$ scattering.

It means that the complicated reaction $a + A \to a + A$ can be replaced by the potential scattering process shown in part (b) of fig. 52. Note that a' is not the original projectile-nucleon a, but an effective particle (or quasi-particle) whose wave function is given by $\rho_E(\vec{k})$. The effective one-body potential U^{opt} corresponds to the empirical optical model potential U^{opt}_{emp} which has played such an important role in explaining the elastic nucleon-nucleus scattering experiments in the past twenty years.

There is, however, an important difference between the above derived potential U^{opt} and the empirical potential U^{opt}_{emp}. The latter is energy-dependent, in the sense that it depends on the asymptotic energy of the projectile a. For example, the empirical optical potential for $p + {}^{16}O$ elastic scattering with $E_p^{lab} = 20$ MeV is different from that with $E_p^{lab} = 40$ MeV. This energy dependence has been a long established practice in nuclear physics[66-68]. On the other hand, the above derived U^{opt} does not have this energy dependence. This can be seen from the diagram of U^{opt} shown in fig. 51 which depends on the single-particle energy $\varepsilon_{\vec{k}''}$ (and thus on the energy spectrum of H_o), but has no dependence on E_p^{lab}. Thus, the folded-diagram theory predicts that there exists an optical potential U^{opt} which can describe, for example, elastic $p + {}^{16}O$ scattering with E_p^{lab} being either 20 MeV or 40 MeV. This is indeed an interesting result. Since the work of Kuo, Osterfeld and Lee[30] was published, several authors[69-71] have further investigated this problem and have, by different approaches, confirmed the existence of a theoretical energy-independent optical model potential. It will thus be of great interest actually to evaluate such optical potentials, using the folded-diagram methods discussed above.

8.2. Free nucleon-nucleon potential. As a second application of the folded-diagram theory to problems other than the shell-model effective interaction, we discuss the derivation of the free nucleon-nucleon potential V_{NN} starting from meson exchanges. It has been a long tradition in nuclear physics that nucleons are taken to interact with each other via a nucleon-nucleon potential. Furthermore, nuclei are considered to be composed of neutrons and protons only, ignoring all the mesonic and other nucleonic degrees of freedom. But in reality, we know that nucleons interact with each other by exchanging elementary particles such as π-mesons. How to define and then derive the nucleon-nucleon potential V_{NN} has been a central problem in nuclear theory for a long time [see, for example, the comprehensive review by Brown and Jackson[38)]]. In short, it may be said that a satisfactory solution to this problem has not yet been found.

To address these questions, Li, Ng and Kuo[31)] considered a generalization of the P- and Q-space concepts which we have discussed repeatedly in the previous sections. Let us consider two possible representations for the nuclear system:

Representation (X+Y) This is the familiar (physical) representation where the nucleus is composed of nucleons, antinucleons, mesons (π, ρ, ω, ...) and other elementary particles. The interactions among the constituent particles are assumed to be described by the usual field theoretical Lagrangians, such as $L_{\pi NN} = igN(\bar{\Psi}\gamma_5\vec{\tau}\cdot\vec{\phi}\psi)$ for the πNN-vertex. (103a)

Representation (X) This is an artificial (effective) representation where the nucleus is composed of neutrons and protons only. The interactions among these particles are described by some operator V_{NN}, which may be referred to as the nucleon-nucleon potential. (103b)

Clearly, (X+Y) corresponds to our familiar complete Hilbert space P+Q, while (X) corresponds to the model space P. In the complete space (X+Y) we have the original interactions $L_{\pi NN}$ and so forth, while in the model space (X) we have the effective interaction V_{NN}. Hence, we must require that

"the physics given by (X) agrees with the physics given by (X+Y)".

This places a requirement on V_{NN}, thus enabling us to determine V_{NN}.

The above principles seem to be both simple and reasonable. The problem to be solved is how to formulate them in a mathematical way. In our derivation of the effective interaction in the previous sections, these principles were formulated in terms of the Schrödinger equations $H\Psi = E\Psi$ and $PH_{eff}P\Psi = EP\Psi$. For systems of strongly interacting elementary particles, it is no longer appropriate to use these equations. One must use covariant field theoretical formulations, as was done by Li, Ng and Kuo[31)]. To outline their work, let us consider a nucleus consisting of two nucleons. The two-nucleon (NN) transition matrix in the (X+Y) representation is denoted by $\langle(NN)_f|T(E_i)|(NN)_i\rangle$, where $(NN)_f$ and $(NN)_i$ are free two-nucleon states and E_i is the energy of the state $(NN)_i$. The effective NN transition matrix in the (X) representation is denoted by $\langle(NN)_f|\bar{T}(E_i)|(NN)_i\rangle$. Clearly, the operator $\bar{T}$ is different from T. But we require that the physics given by $\bar{T}$ agrees with that given by T. Mathematically, this is expressed by

$$\langle(NN)_f|\bar{T}(E_i)|(NN)_i\rangle = \langle(NN)_f|T(E_i)|(NN)_i\rangle , \tag{104}$$

implying that $\bar{T}$ and T have the same NN half-off-shell matrix elements. For given meson-exchange models of the NN interaction, such as those in which two nucleons interact with each other by exchange of pions only, the r.h.s. of eq. (104) is in principle known. The l.h.s. can be expressed in terms of V_{NN}, the effective interaction in the (X) representation, and thus one can in principle solve eq. (104) for V_{NN}. By using the folded-diagram factorization described in the early sections of the present work, Li, Ng and Kuo were able to derive a formally exact and unique expression for V_{NN} from eq. (104).

As there are only neutrons and protons in the (X) representation, it is essential to employ time-ordered Feynman diagrams, in which nucleons and anti-nucleons are treated separately. The nucleons form the basis vectors of the (X) space, while the anti-nucleons belong to the basis vectors of the (Y) space. An irreducible vertex function Σ is defined in a way very similar to our previous two-body $\hat{Q}$-box. Namely, Σ is composed of all the irreducible time-ordered Feynman diagrams, each of which has two incoming and two outgoing nucleon lines. These diagrams have at least two interaction vertices and any interval between two successive vertices must form a (Y) space state. Some examples are given in fig. 53. Then, the solution obtained for V_{NN} is simply the folded-diagram expansion

$$V_{NN} = \Sigma - \Sigma\int\Sigma + \Sigma\int\Sigma\int\Sigma - \cdots \tag{105}$$

This equation has exactly the same structure as the expansions of v_{eff} and U^{opt}, as given by eq. (49) and fig. 50, respectively. Some low-order diagrams of V_{NN} are shown in fig. 54. As compared to earlier theories[38)] of the NN potential, the diagrams $\Pi\int\Pi$ and $\Pi\int\Pi\int\Pi$, etc., are new.[†] Detailed calculation of V_{NN} according to eq. (105) and fig. 53 and further study of its properties are in progress[72)].

For a many-nucleon system, we will in general have a many-body nucleon-nucleon potential V_{NN}^{A}, A being the total number of nucleons. This potential may be defined in a way similar to the two-nucleon potential V_{NN}, namely by the condition

$$\langle(NN\cdots N)_f|\bar{T}(E_i)|(NN\cdots N)_i\rangle = \langle(NN\cdots N)_f|T(E_i)|(NN\cdots N)_i\rangle, \tag{106}$$

† A different folded-diagram theory for V_{NN} has been proposed by Johnson[73)]. As discussed in ref.[31)], there are basic differences between these two theories, both in the definition of V_{NN} and in the folded-diagram methods used for the derivation of V_{NN}. The spirit of the two methods is the same, however, namely to derive an effective NN potential V_{NN} in order to preserve certain essential physical properties of the original system.

where, as before, T is the physical transition matrix and $\overline{T}$ is the effective transition matrix in the (X) space, where the nucleons interact with each other via V_{NN}. Further, E_i is the energy of the free A-nucleon state $(NN\cdots N)_i$. From eq. (106) one can obtain a solution for V^A_{NN} which has exactly the same form as eq. (105), except that all the Σ-boxes have A external nucleon lines, incoming and outgoing. Also, V^A_{NN} will have many-body components and can be analysed in the same way as was done for the many-body effective interaction in subsect. 5.3.

Let us finally remark that our purpose in this subsection has not been to give a detailed derivation of V^A_{NN}, including V_{NN}, but rather to convey the idea that folded diagrams may be a very useful tool for the derivation of the free nucleon-nucleon potential.

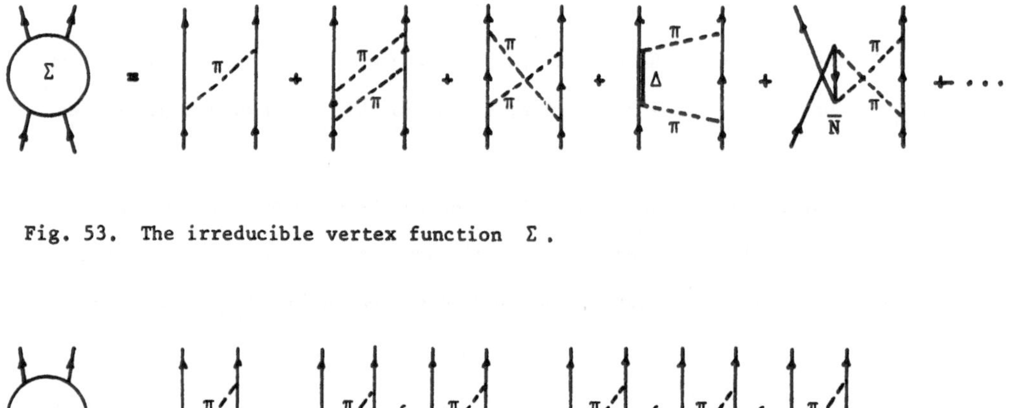

Fig. 53. The irreducible vertex function Σ.

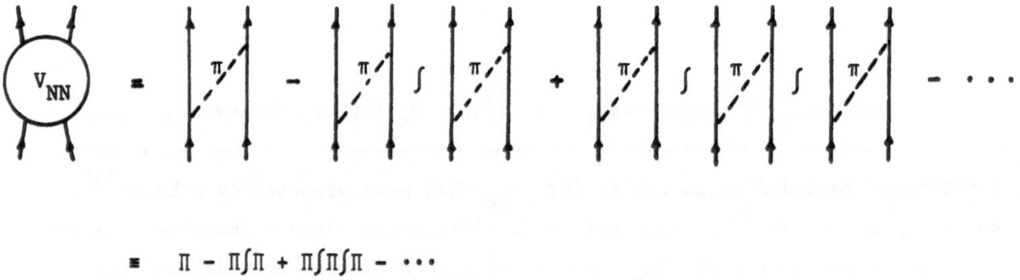

$$\equiv \Pi - \Pi\int\Pi + \Pi\int\Pi\int\Pi - \cdots$$

Fig. 54. Folded-diagram expansion of the two-body nucleon-nucleon potential V_{NN}.

9. Application to atomic and molecular structure calculations

There is a close similarity between the nuclear shell model and the atomic shell model. In fact, it may be said that the former originated from the latter. We have seen in the previous sections that the folded-diagram theory has been very useful for microscopic nuclear shell-model calculations. One may naturally inquire if the folded-diagram theory can also be applied to atomic shell-model calculations. The answer is definitely yes, but to our knowledge relatively few applications have been made.

In comparison to nuclear structure calculations there may indeed be several advantages in applying the folded-diagram theory to atomic structure calculations. Firstly, the inter-electron interaction - the Coulomb force - is well known. (In the nuclear case, the nucleon-nucleon interaction is rather uncertain.) Secondly, the atomic energy levels have been measured extensively and accurately. Thus, it is possible to make a detailed comparison between calculated and measured energy levels. And this comparison may provide an unambiguous test of the many-body method used in the calculation. In contrast, when nuclear structure calculations do not reproduce the measured levels, one cannot be sure about the cause of the deviation. It may be due to defects in the nucleon-nucleon interaction, in the many-body treatment, or in both.

As pointed out in sect. 4, a main function of the folded-diagram theory is to provide a systematic method for reducing the full-space many-body Schrödinger equation $H\Psi_\lambda = E_\lambda\Psi_\lambda$ to a model-space secular equation $PH_{eff}P\Psi_n = E_nP\Psi_n$, where $\{E_n\}$ is a subset of $\{E_\lambda\}$. A crucial point is how to choose the model space. This, of course, requires some knowledge of the physical system to be studied. Thus, in order to apply the folded-diagram method to atomic structure problems, we shall need to review some basics of the atomic shell model[74-76]. Hopefully this may serve as a pedagogical introduction to atomic physics for those readers who, like us, have their experience from nuclear physics. After having summarized the basics of the atomic shell model in sect. 9.1, we shall discuss the application of folded diagrams to structure calculations in open-shell atoms in sect. 9.2. Then, in the final sect. 9.3 we briefly discuss applications of the method to molecular spectroscopy. As we shall see there, the $\hat{Q}$-box formulation of the folded-diagram theory seems to be particularly well suited to deal with molecular structure calculations.

9.1. Basics of the atomic shell model. In atomic physics one commonly chooses units such that

$$m_e = \hbar^2 = e^2/4\pi\varepsilon_o = 1 . \tag{107}$$

With these units, the atomic Hamiltonian for the electrons can be written as, using obvious notations:

$$\begin{aligned} H &= H_o + H_1 , \\ H_o &= \sum_{i=1}^{Z} h_o(i) , \\ h_o(i) &= -\frac{1}{2}\nabla_i^2 - \frac{Z}{r_i} + u(r_i) , \\ H_1 &= \sum_{i>j} \frac{1}{r_{ij}} - \sum_i u(r_i) . \end{aligned} \tag{108}$$

Comparing eq. (108) with the nuclear Hamiltonian, we observe that the atomic Hamiltonian has an external central field $-Z/r_i$ provided by the atomic nucleus, whereas the nuclear Hamiltonian does not have such an external field. In eq. (108), $u(r_i)$ is an auxiliary single-particle potential added to and then subtracted from the Hamiltonian. (Recall the familiar practice of writing $H = T + V = (T + U) + (V - U) \equiv H_o + H_1$ with $U = \sum_i u_i$.) In the nuclear shell model we generally take u_i to be the harmonic oscillator potential. In atomic physics u_i is chosen as the self-consistent Hartree-Fock (HF) field, which has been extensively calculated and tabulated[77]. In fact, standard computer programs have been developed for performing such calculations. The atomic mean field is definitely much better established than the nuclear one. Despite the uncertainty associated with the nucleon-nucleon interaction we feel that more efforts should be put into evaluating the nuclear mean field.

To proceed with atomic structure calculations, we surely need to know the electronic single-particle wave functions ϕ_α and energies ε_α defined by

$$h_o(i)\,\phi_\alpha(r_i) = \varepsilon_\alpha\,\phi_\alpha(r_i)\,. \tag{109}$$

Making the standard separation

$$\phi_\alpha(r) = \frac{P(r)}{r}\,Y^{\ell}_{m_\ell}(\theta,\phi)\,\chi_{m_s}\,, \tag{109.1}$$

where Y is the spherical harmonic and χ the spin wave function, we obtain the following radial Schrödinger equation

$$\{-\frac{1}{2}\frac{d^2}{dr^2} + \frac{\ell(\ell+1)}{2r^2} - \frac{Z}{r} + u(r)\}\,P(r) = \varepsilon\,P(r)\,. \tag{109.2}$$

In eqs. (109.1-2) we have used r instead of r_i. The spectroscopic notations for the orbital angular momentum ℓ are as usual

$$\ell = \begin{cases} 0 & 1 & 2 & 3 & 4 & 5 & 6 & 7 & \dots \\ s & p & d & f & g & h & i & k & \dots \end{cases} \tag{109.3}$$

The principal quantum number is

$$n = \ell + \nu + 1\,, \tag{109.4}$$

where $\nu = 0,1,2,\dots$ is the number of nodes (excluding those at $r = 0$ and $r = \infty$) in the radial wave function $P(r)$. The single-particle energy ε

depends in general on both n and ℓ. For the special case of a pure Coulomb potential (u = 0), ε depends on n only, such as $\varepsilon_n = -13.6/n^2$ eV for Z = 1. In atomic physics the commonly used energy unit is Hartree (H), which is approximately 27.2 eV, i.e. twice the ionization energy for hydrogen.

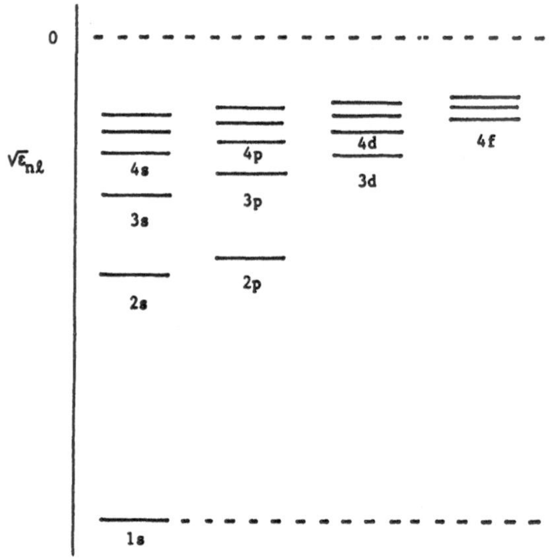

Fig. 55. Schematic level scheme for electronic single-particle orbitals.

In fig. 55 we display a typical level scheme for electronic single-particle orbitals. We note that it is rather different from the level scheme of the single-particle nuclear shell model. The spacings between the major shells are in general relatively smaller than in the nuclear case. This may make it difficult to choose a model space P which is well separated from the excluded Q-space. When this cannot be done, it will be necessary to sum diagrams to high order to obtain convergence of the effective interaction, and this may make the $\hat{Q}$-box formulation of the folded-diagram method particularly useful for atomic structure calculations.

Having established the single-electron wave functions and energies, we proceed to evaluate the matrix elements of the Coulomb interaction between a pair of electrons. The Coulomb interaction is expressed in a standard way as[78)]

$$\frac{1}{r_{12}} = \sum_k \frac{r_<^k}{r_>^{k+1}} \underset{\sim}{C}^k(1) \cdot \underset{\sim}{C}^k(2) , \tag{110}$$

with

$$r_> = \max (r_1, r_2) ,$$

$$r_< = \min (r_1, r_2) ,$$

$$C_q^k = \sqrt{\frac{4\pi}{2k+1}} \; Y_q^k(\hat{r}_i) , \quad i = 1,2 . \tag{110.1}$$

The scalar product of spherical tensors is defined as

$$\underset{\sim}{C}^k \cdot \underset{\sim}{C}^k = \sum_q (-1)^q \, C_q^k \, C_{-q}^k . \tag{110.2}$$

The angular momenta of the electrons may be coupled in the LS or jj representations. The corresponding two-electron wave functions are written as

$$|(n\ell)_1 (n\ell)_2 \; SM_S \; LM_L> \qquad \text{(LS coupling)} ,$$

$$|(n\ell j)_1 (n\ell j)_2 \; JM> \qquad \text{(jj coupling)} . \tag{111}$$

The matrix elements of the Coulomb potential between such states are calculated using standard Racah algebra. We have for instance in LS coupling

$$< (n\ell)_a (n\ell)_b \; SM_S \; LM_L | r_{12}^{-1} | (n\ell)_c (n\ell)_d \; SM_S \; LM_L>$$

$$= \sum_k R^k(ab,cd) \; (-1)^{\ell_c+\ell_b+L} \begin{Bmatrix} \ell_a & \ell_b & L \\ \ell_d & \ell_c & k \end{Bmatrix} < \ell_a \| \underset{\sim}{C}^k \| \ell_c >< \ell_b \| \underset{\sim}{C}^k \| \ell_d > , \tag{112}$$

where the Slater radial integral is

$$R^k(ab,cd) = \iint dr_1 dr_2 \; P_a^*(r_1) P_b^*(r_2) \frac{r_<^k}{r_>^{k+1}} P_c(r_1) P_d(r_2) . \tag{112.1}$$

In eq. (112), $\{ \}$ is the 6j-symbol and $\langle \| \; \| \rangle$ is the standard reduced matrix element, namely

$$\langle \ell \| \underset{\sim}{C}^{k} \| \ell' \rangle = (-1)^{\ell} \, \hat{\ell} \, \hat{\ell}' \begin{pmatrix} \ell & k & \ell' \\ 0 & 0 & 0 \end{pmatrix}, \tag{112.2}$$

where $\hat{x} = \sqrt{2x+1}$ and $(\;)$ is the 3j-symbol. The jj-coupled matrix element of r_{12}^{-1} is calculated in a similar way[74].

The above procedure is in fact not unfamiliar to nuclear physics applications, as essentially the same basic technique is used to evaluate the corresponding nuclear matrix elements. In many ways the calculation of nuclear matrix elements is more complicated. There we need to perform the Brody-Moshinsky transformation between the laboratory (r_1, r_2) and relative and center-of-mass (r, R) coordinates and then evaluate the Talmi integrals for the nucleon-nucleon interaction[79]. When using realistic nucleon-nucleon interactions, we also need to calculate the nuclear reaction matrix[44], as pointed out in sect. 7. This is indeed a rather cumbersome undertaking. Thus, it seems that atomic structure calculations can be carried out more rigorously as well as more conveniently than nuclear structure calculations.

<u>9.2. Open-shell atoms.</u> We are now prepared to discuss how the folded-diagram theory can be applied to the calculation of the structure of open-shell atoms[74,80,81]. A classroom example of nuclear structure calculations is ^{18}O, which has two valence nucleons in the 1s0d shell outside a closed-shell ^{16}O core. We may use the beryllium atom (Be) as a corresponding example of open-shell atomic structure calculations. Referring to fig. 55, we see that the most likely electronic configuration for the ground state of Be is $(1s)^2(2s)^2$. But since the 2p orbit is close by, the admixture of this orbit into the low-lying states of Be may be important and should be taken care of. (In nuclear shell-model language the neglect of the 2p orbit in Be is similar to the exclusion of the $1s_{1/2}$ and $0d_{3/2}$ orbits in ^{18}O

calculations.) Thus, an appropriate model space for the low-lying states of the Be atom would be $(1s)^2(2s,2p)^2$, i.e. a closed $(1s)^2$ core with two valence electrons allowed to wander in the entire 2s2p shell.

The diagrammatic notations used in atomic physics are slightly different from those used for the nuclear case in the previous sections. In fig. 56 we use the example of Be to illustrate the difference. In ref. [82] the electronic orbits are classified into three categories - the virtual, valence and core orbits. In the terminology used for the nuclear case above (see e.g. the caption to fig. 1) they would be denoted as passive-particle, active-particle and passive-hole orbits, respectively. For the core orbits the notations used in figs. 56 and 1 are the same. For valence orbits a line with double arrows is used in atomic calculations, while a single arrow was used for the nuclear case in the previous sections. The graphical notations for the virtual orbits are also different, a clean line with a single arrow being used for the atomic case and a railed line with arrow for the nuclear case.

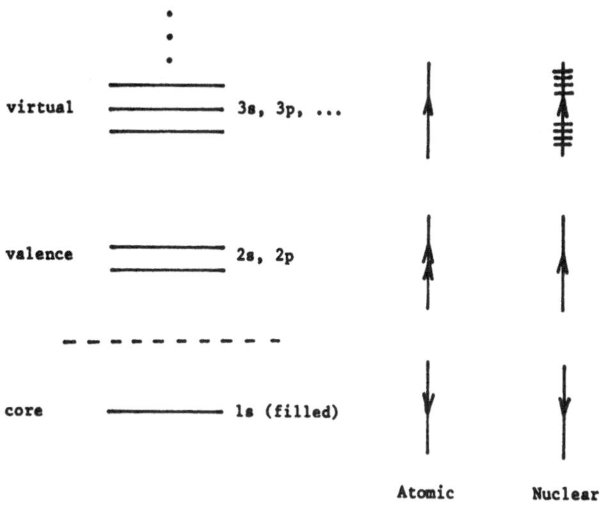

Fig. 56. Classification of shell-model single-particle orbits for the calculation of the low-energy spectrum of the Be atom. The notations for the corresponding particle lines are compared for the atomic and nuclear cases to the right.

bare interaction

core polarization

folded diagram

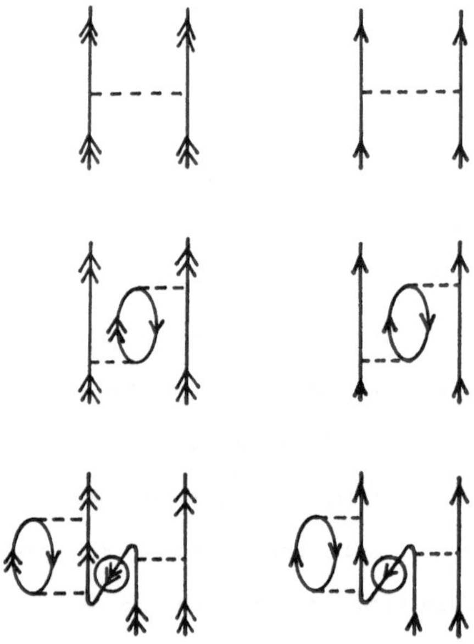

Fig. 57. Diagrams contributing to the shell-model effective interaction in atoms and nuclei.

Clearly, the methods we have described in the previous sections are readily applicable to open-shell atomic calculations. We only need to make some relatively simple changes in the diagrammatic notation. In fig. 57 we have drawn some representative diagrams for the model-space effective interaction. Aside from some nominal notational differences the diagrams contributing to the P-space effective interactions for open-shell atoms and open-shell nuclei are identical. In passing we note that the folded diagrams are often referred to as "backwards" diagrams in atomic physics literature[74,82].

Salomonson et al.[83] have calculated the model-space effective interaction for the valence electrons of Be. Although they did not follow the $\hat{Q}$-box formulation of the folded-diagram theory discussed in previous sections but evaluated the contributions order by order in the Coulomb interaction, it is of interest to study their results in some detail.

For the ^1S (spin singlet and $L = 0$) states of Be they used a two-dimensional model space with basis vectors

$$|\Phi_1\rangle = |(2s)^2;\ ^1S\rangle , \qquad |\Phi_2\rangle = |(2p)^2;\ ^1S\rangle . \tag{113}$$

The effective Hamiltonian is written as

$$H_{eff} = H_o + v^{(1)}_{2\text{-body}} + v^{(2)}_{2\text{-body}} + \cdots \tag{114}$$

where the superscript denotes the order of the Coulomb interaction. Salomonson et al. used an unperturbed Hamiltonian H_o including the $(1s)^2$ HF potential, the single-electron energies so obtained being

$$\varepsilon_{1s} = -5.66717 \text{ H},$$

$$\varepsilon_{2s} = -0.66609 \text{ H},$$

$$\varepsilon_{2p} = -0.51945 \text{ H}. \tag{114.1}$$

We note in passing that atomic structure calculations are carried out to a much higher degree of accuracy than the nuclear ones. Excluding the constant energy contribution from the $(1s)^2$ core electrons, Salomonson et al. obtain the following energy matrix to diagonalize in the two-dimensional space of eq. (113):

	Φ_1	Φ_2		
Φ_1	-1.33218	0	←	H_o
	+0.3964	-0.1438	←	$v^{(1)}_{2\text{-body}}$
	-0.0434	+0.0757	←	$v^{(2)}_{2\text{-body}}$
Φ_2	0	-1.03890		
	-0.1438	+0.4565		
	+0.0437	-0.1237		

(114.2)

All energies are here in units of Hartree (H).

These results are quite instructive. Compared to the splitting between the 2s and 2p orbits the residual effective interaction is in fact quite strong. This means that it is necessary to include the 2p orbit in the model space, as anticipated above. The ground-state wave function obtained by Salomonson et al., including both first and second order effective interactions, is

$$|\Psi_o\rangle = 0.947\ |(2s)^2;\ ^1S\rangle + 0.322\ |(2p)^2;\ ^1S\rangle . \tag{114.3}$$

The $(2p)^2$ component is important and the amount of configurational admixture is approximately the same as in typical nuclear shell-model wave functions.

How do these calculations agree with experiment? The ground-state energy[84)] of the Be atom is -14.6674 H. The result given by the above calculation, including the first and second order effective interactions and also the $(1s)^2$ core contribution, is -14.661 H. Such an agreement is surely astounding compared to the typical agreement obtained in nuclear structure calculations. One may ask if this implies that all the numerous higher order terms not considered in the calculation are unimportant. Although the relative magnitudes of the contributions from the first and second order shown in the matrix (114.2) do indicate smooth convergence, one cannot be sure of this until higher order terms have been evaluated as well. Further studies along this line would thus be desirable.

It may be instructive to discuss in some detail how the calculation of Salomonson et al. relates to the methods described for the nuclear case in the previous sections. First, we point out that Salomonson et al. used a wave operator approach[33,55,56,74)]. Recall from eq. (31.1) that the effective Hamiltonian can be written as

$$H_{eff} = P(H_o + V\Omega)P , \tag{115}$$

where Ω is the wave operator and $V\Omega$ is the effective interaction. One expands Ω as

$$\Omega = 1 + \Omega^{(1)} + \Omega^{(2)} + \cdots \tag{115.1}$$

$\Omega^{(n)}$ denoting the wave operator n-th order in the interaction. One derives $\Omega^{(n)}$ from the iterative equation

$$[\Omega, H_o]P = V\Omega P - \Omega P V \Omega P . \tag{115.2}$$

The n-th order effective interaction is given by $V\Omega^{(n-1)}$. The main difference between this approach and our folded-diagram approach concerns the way of grouping the various terms of the effective interaction series. Here, the terms are grouped according to the power of the bare interaction. (In fact, since Salomonson et al. terminate the series at second order in the Coulomb interaction, they do not include any folded diagrams.) On the other hand, in our $\hat{Q}$-box folded-diagram approach we do not perform an order-by-order summation of the terms. There, we first choose the diagrams to be included in the $\hat{Q}$-box. For example, we may take $\hat{Q}$ to be composed of all the relevant first and second order diagrams. Once $\hat{Q}$ is chosen, we evaluate the folded terms $-\hat{Q}\int\hat{Q}$, $+\hat{Q}\int\hat{Q}\int\hat{Q}$, etc. Thus, we are grouping the terms of the effective interaction according to the number of folds. In each group there will then be terms of various orders in the bare interaction. It would be of interest to calculate the effective interaction for open-shell atoms using the $\hat{Q}$-box formulation of the folded-diagram theory.

The Be atom with two valence electrons is at the beginning of the 2s2p shell. From the nuclear shell model we know that calculations for nuclei in the middle of a major shell are much more complicated than for nuclei at the beginning or the end of the shell. The situation is similar for atoms in the middle of a shell. As an example of such calculations we mention the calculation for carbon by Morrison and Salomonson[81], with some low-order folded diagrams included.

Atomic spectroscopic data have been well tabulated [see e.g. ref.[85]]. In table 4 are listed the ionization potentials (the least energy needed to pull out an electron) and the unperturbed ground-state configurations for the 2s2p shell atoms. Note that the Ne atom has a particularly large ionization potential, reflecting the closed-shell effect. In conclusion, it seems that many of the methods developed for nuclear structure calculations, such as the various effective interaction theories and large shell-model calculations[86], can be applied to open-shell atomic calculations. However, there have not been many effective interaction calculations for mid-shell atoms.

Table 4. Ionization potentials (IP) and unperturbed ground-state configurations for the 2s2p shell atoms[85].

Z	Element	IP [eV]	Number of electrons in orbitals 1s	2s	2p
3	Li	5.392	2	1	0
4	Be	9.322	2	2	0
5	B	8.298	2	2	1
6	C	11.260	2	2	2
7	N	14.534	2	2	3
8	O	13.618	2	2	4
9	F	17.422	2	2	5
10	Ne	21.564	2	2	6

9.3. Molecular spectroscopy. There are a number of similarities between spherical nuclear shell-model calculations and open-shell atomic structure calculations, as we just have seen. Proceeding then to molecular spectroscopic calculations, we find that these may be compared to deformed nuclear shell-model calculations. The electrons in a molecule are subject to a

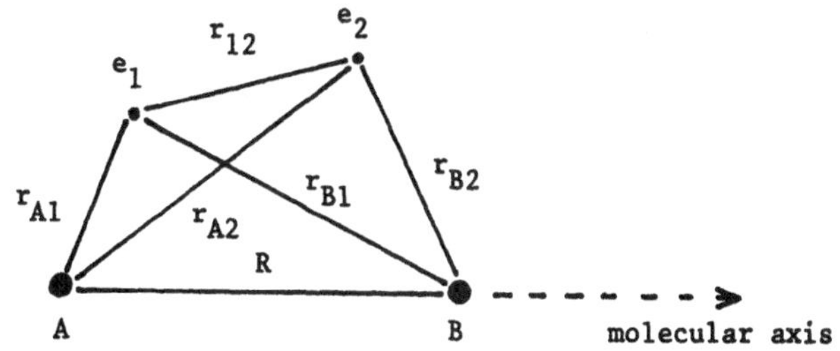

Fig. 58. Schematic picture of H_2 molecule with definition of coordinates.

non-spherical multicenter Coulomb field provided by the nuclei of the molecule. For instance, the Coulomb field of the H_2 molecule is two-centered, provided by the two protons separated by a distance commonly denoted by R.

Let us study this molecule in some detail, in order to review the basics of molecular structure calculations as well as to see how the folded-diagram method can be a useful tool for molecular studies. As shown in fig. 58, we denote the electrons of this molecule by e_1 and e_2 and the two nuclei (protons) by A and B. The molecular Hamiltonian can be written as

$$H = H_{el} + H_{nuc} ,$$
$$H_{el} = -\frac{1}{2}(\nabla_1^2 + \nabla_2^2) - \frac{e^2}{r_{A1}} - \frac{e^2}{r_{B1}} - \frac{e^2}{r_{A2}} - \frac{e^2}{r_{B2}} + \frac{e^2}{r_{12}} ,$$
$$H_{nuc} = -\frac{\hbar^2}{2m_N}(\nabla_A^2 + \nabla_B^2) + \frac{e^2}{R} . \qquad (116)$$

Our goal is eventually to solve the Schrödinger equation

$$H\,\Psi(r,R) = E_{mol}\,\Psi(r,R) , \qquad (117)$$

where r denotes the electronic coordinates and R the nuclear coordinates.

(The subscripts el, nuc and mol denote respectively electronic, nuclear and molecular.) It is a common practice to employ the Born-Oppenheimer approximation in separating the wave function as

$$\Psi(r,R) = \Psi_{el}(r,R)\,\Psi_{nuc}(R)\,. \tag{117.1}$$

Furthermore, one treats the nuclei as being fixed at a separation R from each other. Then, we have the reduced problem

$$H_{el}\,\Psi_{el} = E_{el}\,\Psi_{el}\,, \tag{117.2}$$

$$E_{mol} = E_{el} + \frac{e^2}{R}\,. \tag{117.3}$$

In this way the molecular energy is calculated as a function of R, and the equilibrium separation R_e for the molecule is given by $E_{mol}(R_e) = \min$.

Thus, the main task is now to solve the electronic many-body equation (117.2). Historically there are two ways to proceed, and they differ from each other mainly in how to divide H_{el} into the unperturbed part H_o and the interaction part H_{int}. In the Heitler-London or valence-bond treatment one takes

$$H_o' = -\frac{1}{2}(\nabla_1^2 + \nabla_2^2) - \frac{e^2}{r_{A1}} - \frac{e^2}{r_{B2}}\,,$$

$$H_{int}' = -\frac{e^2}{r_{B1}} - \frac{e^2}{r_{A2}} + \frac{e^2}{r_{12}}\,, \tag{118}$$

with $H_{el} = H_o' + H_{int}'$. Apparently this scheme does not seem to be suitable for many-body perturbation calculations as H_o' is considerably different from H_{el}. (In fact, H_o' corresponds to H_{el} at very large separation R.) A more commonly used method is the Hund-Mulliken or molecular-orbital (MO) approach, where one writes

$$H_o = \sum_{i=1,2} h(i) ,$$

$$h(i) = -\frac{1}{2}\nabla_i^2 - \frac{e^2}{r_{Ai}} - \frac{e^2}{r_{Bi}} ,$$

$$H_{int} = \frac{e^2}{r_{12}} , \tag{119}$$

again with $H_{el} = H_o + H_{int}$. It is straightforward to generalize the above to more complicated molecules. The MO approach is a standard method commonly used in molecular structure calculations[87,88].

In order to apply the folded-diagram method to molecular problems, one first has to solve the single-electron problem

$$h\,\phi_\alpha = \varepsilon_\alpha\,\phi_\alpha \ . \tag{120}$$

The eigenfunctions Φ_n of H_o are just Slater determinants composed of the ϕ_α's. We may choose a model space $P = \sum_{i=1}^{d} |\Phi_i\rangle\langle\Phi_i|$. Then, the molecular Schrödinger equation can be reduced to a model-space secular equation

$$P\,H_{el}^{eff}\,P\,\Psi_{el} = E_{el}\,P\,\Psi_{el} , \tag{121}$$

where H_{el}^{eff} may be obtained in a systematic way using the folded-diagram method.

The wave functions and energies of the molecular orbitals are given by the solution of eq. (120). And we need to calculate them first in order to solve eq. (121). This calculation is, however, much more complicated than the corresponding atomic calculation, because now the Coulomb field contained in h is non-spherical. Referring to fig. 58, the molecule is rotationally invariant with respect to the molecular axis z. Hence, m (the z-component of the orbital angular momentum) is a good quantum number for the molecular orbitals. In fact, the quantum numbers usually used in labelling the molecular orbitals for diatomic molecules are

$$m = 0, \pm 1, \pm 2, \pm 3, \pm 4, \ldots$$

$$\lambda = \begin{cases} 0, & 1, & 2, & 3, & 4, & \ldots \\ \sigma, & \pi, & \delta, & \phi, & \gamma, & \ldots \text{ (code)} , \end{cases} \tag{122}$$

where λ is defined as $|m|$ and is the number of nodal planes, containing the molecular axis, of the respective molecular orbitals. For several electrons the z-component of the total orbital angular momentum is a good quantum number and we use the following notation for labelling these states:

$$M = m_1 + m_2 + m_3 + \cdots$$

$$\Lambda = |M| = \begin{cases} 0, 1, 2, 3, 4, \ldots \\ \Sigma, \Pi, \Delta, \Phi, \Gamma, \ldots \text{ (code)} . \end{cases} \tag{123}$$

Numerical solution of eq. (120) is typically carried out by approximating the ϕ_α's as linear combinations of a finite number of convenient basis functions. For diatomic (and linear polyatomic) molecules the basis set usually consists of the Slater type orbitals (STO) centered on each of the nuclei in the molecule. For non-linear polyatomic molecules it is a common practice to use basis functions consisting of Gaussian type orbitals (GTO), also centered on each nucleus of the molecule. The STO and GTO differ from each other mainly in their radial wave functions; the former has $r^{n-1}e^{-\zeta r/a_o}$ whereas the latter has $r^{\ell}e^{-\zeta r^2/a_o^2}$. The orbitals of more complicated molecules are classified using group-theoretical methods[88).

As indicated by fig. 59, the molecular orbitals approach degenerate atomic orbitals in the limit of large nuclear separation. This clearly tells us that the spectrum of molecular orbitals is generally more densely packed than that of atomic orbitals, and the energy gaps between major shells are thus less pronounced. This explains why the folded-diagram method may be well suited for molecular calculations. Because of the high density of single-electron orbitals in a molecule, one can only afford to include a

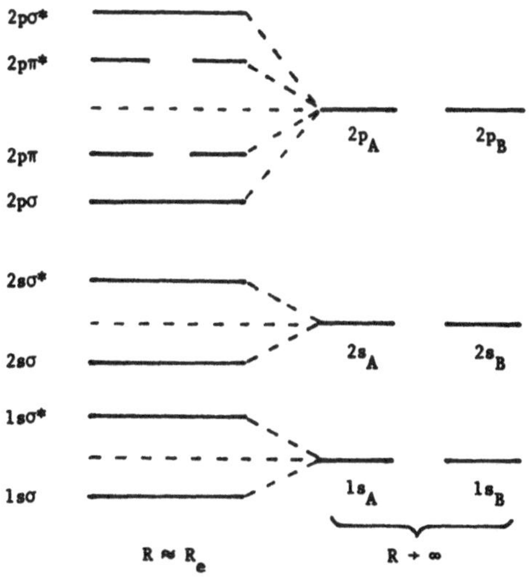

Fig. 59. Evolution of molecular orbitals with respect to nuclear separation[87]. The states on the left are denoted by $n\ell\lambda$, where n and ℓ are respectively the principle and orbital angular momentum quantum numbers of the corresponding state in the united atom ($R \to 0$). Bonding orbitals (i.e. increased binding at R_e compared to $R \to \infty$) are denoted by σ , π , ... and anti-bonding orbitals (i.e. decreased binding at R_e compared to $R \to \infty$) by σ^*, π^*, ...

small number of them in the P-space in which the Hamiltonian matrix is set up and diagonalized. Thus, we must use an effective interaction $PV_{eff}P$ which is different from the Coulomb interaction. As the Q-space is generally not well separated from the P-space, we may have to include diagrams to high order to obtain reasonable convergence. This speaks for performing partial summations similar to those done in the $\hat{Q}$-box formulation of the folded-diagram theory.

In atomic and molecular physics such theories are usually referred to as multi-configurational many-body perturbation theory (MCMBPT). Kaldor[89] is probably the first one to apply the MCMBPT approach to molecular spectroscopy, evaluating the lowest four excited Σ^+ states of H_2 at a nuclear separation of $R = 1.4$ Bohr with the inclusion of the third order once-folded diagrams. Similar folded-diagram calculations were later performed by Stern and Kaldor[90] for the ground and seven excited states of the BH molecule. Yamamoto and

Saika[32] have performed MCMBPT calculations of the H_2O, C_2H_4 and H_2 molecules with the folded diagrams evaluated using the Q-box approach described in sects. 5.4 and 7.

It should be informative and of interest to list some of the results of Yamamoto and Saika. They have calculated ionization potentials (IP), electronic affinities (EA) and excitation energies (EE). [IP is the electron separation energy for the hole molecular orbitals, while EA denotes -IP for the particle orbitals.] Some representative results of their calculation are listed in table 5.

Table 5. Results from the MCMBPT calculation of Yamamoto and Saika[32].

Molecule	Hole species	IP_o	IP_2	IP_3	IP_F	Total	Expt.
H_2O	$1a_1$	20.558	-1.081	0.715		20.192	19.835
H_2	$1\sigma_g$	0.5946	0.0058	0.0017	0.0001	0.6029	(0.6050)
	Particle species	EA_o	EA_2	EA_3	EA_F	Total	Expt.
H_2O	$4a_1$	0.229	-0.027	0.001		0.203	
H_2	$2\sigma_g$	-0.0295	0.0007	0.0001	0.0000	-0.0287	

All entries are in atomic units (Hartree). For H_2O, the states denoted by a_1 are fully symmetric under a rotation of 2π about the z-axis and reflection about a plane perpendicular to the z-axis. Thus, a state na_1 is the n-th lowest state of symmetry a_1. For H_2, the states $1\sigma_g$ and $2\sigma_g$ are the first and second $\lambda = 0$ states with even parity under coordinate inversion. For further details on the classification of states, see e.g. ref.[88].

In the table IP_o and EA_o are the zeroth order (i.e. HF) results. The first order results IP_1 and EA_1 are zero in the HF approximation. Then, IP_2, EA_2, IP_3 and EA_3 are the second and third order corrections. In the H_2O calculations the authors used GTO basis functions. As shown, the order-by-order convergence is excellent for both IP and EA. And the total result is also in good agreement with the experimental value for the $1a_1$ state. For the H_2O case, the authors have not listed the folded-diagram contributions IP_F and EA_F. For the H_2 molecule, the order-by-order convergence is again excellent for both IP and EA. Here, the folded-diagram contributions are negligibly small. In fact, for the case shown they found that the once-folded contribution to IP is $IP_{F1} = 0.0001105$ while the twice-folded one is $IP_{F2} = 0.0000004$. No experimental value was available for the IP of the $1\sigma_g$ state and the authors listed an accurate theoretical value (which presumably is rather close to the experimental one) in parentheses.

We are indeed amazed by the good results of the molecular many-body calculations, in terms of the agreement with experiment, the smallness of the higher order corrections as well as the smallness of the folded-diagram corrections. We do not really know the reason for this. However, since the Coulomb interaction is well known, one may take the optimistic point of view that the many-body methods used in the calculation are an excellent tool for such calculations. But such a conclusion would obviously have to be tested by further, systematic studies of the $\hat{Q}$-box and related folded-diagram methods. This would, in fact, be very worthwhile, as theoretical studies of molecular spectroscopy are a topic of much current interest[80,91].

10. Conclusions

In this paper we have given a rather comprehensive review of the derivation of a valence-linked energy-independent effective interaction for a general degenerate model space. Using time-dependent perturbation theory, folded diagrams enter in a natural way into the perturbation expansion of this effective Hamiltonian. The cancellation of the disconnected diagrams is shown in a simple and rigorous way.

A noteworthy feature of this effective Hamiltonian is its close similarity to the highly successful empirical shell-model effective Hamiltonian. Thus, the present approach provides a microscopic foundation of the empirical shell-model approach. For example, the theory described in this paper allows us to make use of the experimental information on ^{16}O and ^{17}O to calculate properties of ^{18}O. In this calculation we only need to evaluate the two-body effective interaction, which may be compared to the empirical shell-model effective interaction. For nuclei with more than two valence nucleons, the present method gives a rigorous derivation of the many-body effective interaction as a sum of two-, three-, ... and N_V-body terms, N_V being the number of valence nucleons. Furthermore, the n-body term is the same for all nuclei with $N_V \geq n$. For example, for ^{19}O, v_{eff} consists of a two-body term $v_{eff}(2)$, which is the same as for ^{18}O, and a three-body term $v_{eff}(3)$.

Detailed rules for the calculation of the folded-diagram expansion of the effective Hamiltonian have been given. In fact, the calculation of each term in the expansion is in principle straightforward. However, in actual calculations we do not have to evaluate individual folded diagrams. We only need to calculate the non-folded diagrams and their energy derivatives, and this makes the application of the folded-diagram method very simple.

Thus, we are now able to perform a systematic first-principle calculation of nuclear properties, starting from the free nucleon-nucleon interaction. Two remarks should be made in this connection. First, it is necessary to take the strong short-range correlations of the nucleon-nucleon interaction into account by G-matrix partial summation. Secondly, we need to make truncations in the calculation of the so-called $\hat{Q}$-box, which is defined as the sum of all the non-folded diagrams. Two kinds of truncations have to be made. In each individual $\hat{Q}$-box diagram we have to truncate the intermediate-state summation. The use of a nucleon-nucleon interaction with a weak tensor force component seems to expedite the convergence of this summation. Further, we need to restrict the number of diagrams which are included in the $\hat{Q}$-box. This truncation may be facilitated by choosing a good single-particle basis, obtained for example by the Hartree-Fock (or Brueckner-Hartree-Fock) approximation. There are indications, however, that the truncation of the $\hat{Q}$-box series is not so critical, provided that folded diagrams be included.

Finally, we emphasize that the present folded-diagram theory is very general in structure and thus applicable to several other physical problems. As further examples of the versatility of this theory we have sketched the derivations of the optical model potential for elastic scattering and of the free nucleon-nucleon potential from meson exchange. In the final section we have discussed the applicability of the folded-diagram theory to atomic and molecular structure problems. In fact, there is close similarity between the atomic shell model and the spherical nuclear shell model, while molecular models show similarities with the non-spherical nuclear shell model. Furthermore, the density of single-particle orbits is in general relatively higher in atoms and molecules than in nuclei, so that one will have to work in a fairly restricted model space P not well separated from the excluded space Q. Thus, one may have to include several higher order diagrams to obtain a reasonable effective interaction. This may serve to make the $\hat{Q}$-box formulation of the folded-diagram theory a powerful tool for atomic and molecular structure calculations. This method has in fact been applied to calculations of the H_2 and H_2O molecules with very promising results[32)].

Application of the folded-diagram theory to atomic and molecular structure problems also has important methodological aspects. Since the inter-electron interaction is well known and a large number of atomic and molecular levels have been measured with great precision, it should be possible to use atoms and molecules to study the validity of the folded-diagram approach to quantal many-body problems.

References

1. D.J. Thouless, The quantum mechanics of many-body systems (Academic, New York, 1961)
2. G.E. Brown, Many-body problems (North-Holland, Amsterdam, 1972)
3. A.L. Fetter and J.D. Walecka, Quantum theory of many-particle systems (McGraw-Hill, New York, 1971)
4. H. Kümmel, K.H. Lührmann and J.G. Zabolitzky, Phys. Rep. 36, No. 1 (1978) 1
5. J. Speth, E. Werner and F. Wild, Phys. Rep. 33, No. 3 (1977) 127
6. C.A. Engelbrecht, F.J.W. Hahne and W.D. Heiss, Ann. Phys. (N.Y.) 104 (1977) 221;
   F.J.W. Hahne, W.D. Heiss and C.A. Engelbrecht, Ann. Phys. (N.Y.) 104 (1977) 251;
   W.D. Heiss, C.A. Engelbrecht and F.J.W. Hahne, Ann. Phys. (N.Y.) 104 (1977) 274
7. D.R. Bés, R.A. Broglia, G.G. Dussel, R.J. Liotta, H.M. Sofía and R.J. Perazzo, Nucl. Phys. A260 (1976) 1; 27; 77 ;
   P.F. Bortignon, R.A. Broglia, D.R. Bés and R. Liotta, Phys. Rep. 30, No. 4 (1977) 305
8. J.P. Elliott and A.M. Lane, in Encyclopedia of physics, ed. S. Flügge, (Springer, Berlin, 1957) p. 241
9. I. Talmi, Rev. Mod. Phys. 34 (1962) 704
10. B.R. Barrett and M.W. Kirson, in Advances in nuclear physics, ed. M. Baranger and E. Vogt, vol. 6 (Plenum, New York, 1973) p. 219
11. T.T.S. Kuo, Ann. Rev. Nucl. Sci. 24 (1974) 101
12. P.J. Ellis and E. Osnes, Rev. Mod. Phys. 49 (1977) 777
13. T.T.S. Kuo and G.E. Brown, Nucl. Phys. 85 (1966) 40
14. T.T.S. Kuo and G.E. Brown, Nucl. Phys. A114 (1968) 241
15. B.R. Barrett and M.W. Kirson, Nucl. Phys. A148 (1970) 145
16. M.W. Kirson, Ann. Phys. (N.Y.) 66 (1971) 624; 82 (1974) 345
17. T H. Schucan and H.A. Weidenmüller, Ann. Phys. (N.Y.) 73 (1972) 108; 76 (1973) 483

18. P.J. Ellis and E. Osnes, Phys. Lett. 45B (1973) 425;
J.M. Leinaas and E. Osnes, Phys. Scr. 22 (1980) 193

19. M.R. Anastasio, T.T.S. Kuo, T. Engeland and E. Osnes, Nucl. Phys. A271 (1976) 109

20. J.M. Leinaas and T.T.S. Kuo, Ann. Phys. (N.Y.) 98 (1976) 177;
Phys. Lett. 62B (1976) 275

21. K. Andō and H. Bandō, Prog. Theor. Phys. 52 (1975) 1711;
K. Andō, H. Bandō and S. Nagata, Prog. Theor. Phys. 57 (1977) 1303; 1584

22. T. Morita, Prog. Theor. Phys. 29 (1963) 351

23. G. Oberlechner, F. Owono-N'-Guema and J. Richert, Nuovo Cim. B68 (1970) 23

24. M.B. Johnson and M. Baranger, Ann. Phys. (N.Y.) 62 (1971) 172

25. T.T.S. Kuo, S.Y. Lee and K.F. Ratcliff, Nucl. Phys. A176 (1971) 65;
K.F. Ratcliff, S.Y. Lee and T.T.S. Kuo, unpublished

26. B.H. Brandow, Rev. Mod. Phys. 39 (1967) 771

27. I. Lindgren, J. Phys. B7 (1974) 2441

28. T.T.S. Kuo, in Dynamic structure of nuclear states, ed. D.J. Rowe, L.E.H. Trainor and T.W. Donnelly (University of Toronto, Toronto, 1972) p. 205

29. K.F. Ratcliff, in Effective interactions and operators in nuclei, ed. B.R. Barrett, Lecture notes in physics, vol. 40 (Springer, Berlin, 1975) p. 42

30. T.T.S. Kuo, F. Osterfeld and S.Y. Lee, Phys. Rev. Lett. 45 (1980) 786

31. Guang-lie Li, K.K. Ng and T.T.S. Kuo, Phys. Rev. C25 (1982) 3043

32. S. Yamamoto and A. Saika, Chem. Phys. Lett. 78 (1981) 316

33. C. Bloch, Nucl. Phys. 6 (1958) 329; 7 (1958) 451;
C. Bloch and J. Horowitz, Nucl. Phys. 8 (1958) 91

34. M. Gell-Mann and F. Low, Phys. Rev. 84 (1951) 350

35. H.J. Lipkin, N. Meshkov and A.J. Glick, Nucl. Phys. 62 (1965) 188

36. H.A. Bethe, B.H. Brandow and A.G. Petschek, Phys. Rev. 129 (1963) 225

37. J. Goldstone, Proc. Roy. Soc. (London) A293 (1957) 267

38. G.E. Brown and A.D. Jackson, The nucleon-nucleon interaction (North-Holland, Amsterdam, 1976)

39. R.V. Reid, Ann. Phys. (N.Y.) 50 (1968) 411

40. K. Holinde, R. Machleidt, M.R. Anastasio, A. Faessler and H. Müther, Phys. Rev. C18 (1978) 18

41. M. Lacombe, B. Loiseau, J.M. Richard, R.V. Mau, J. Cote, P. Pires and R.D. Tourreil, Phys. Rev. C21 (1980) 861, and references quoted therein

42. M. Harvey, Nucl. Phys. A352 (1981) 301; 326

43. K. Holinde, Phys. Rep. 68, No. 3 (1981) 121

44. E.M. Krenciglowa, C.L. Kung, T.T.S. Kuo and E. Osnes, Phys. Lett. 63B (1976) 141; Ann. Phys. (N.Y.) 101 (1976) 154

45. J. Shurpin, D. Strottman, T.T.S. Kuo, M. Conze and P. Manakos, Phys. Lett. 69B (1977) 395

46. D. Grillot and H. McManus, Nucl. Phys. A113 (1968) 161

47. B.R. Barrett, R.G.L. Hewitt and R.J. McCarthy, Phys. Rev. C3 (1971) 1137

48. J.P. Vary and S.N. Yang, Phys. Rev. C15 (1977) 1545

49. S.F. Tsai and T.T.S. Kuo, Phys. Lett. 39B (1972) 427

50. M. Sommermann, private communication

51. E.M. Krenciglowa, T.T.S. Kuo, E. Osnes and B. Giraud, Phys. Lett. 47B (1973) 322

52. W. Chung, Ph.D. thesis (Michigan State University, 1976, unpublished); B.H. Wildenthal and W. Chung, in Mesons and nuclei, ed. M. Rho and D. Wilkinson, vol. 2 (North-Holland, Amsterdam, 1979), p. 721

53. T.T.S. Kuo, J. Shurpin, K.C. Tam, E. Osnes and P.J. Ellis, Ann. Phys. (N.Y.) 132 (1981) 237

54. E.M. Krenciglowa and T.T.S. Kuo, Nucl. Phys. A235 (1974) 171

55. S.Y. Lee and K. Suzuki, Phys. Lett. 91B (1980) 173; Prog. Theor. Phys. 64 (1980) 2091

56. T.T.S. Kuo, in Topics in nuclear physics, ed. T.T.S. Kuo and S.S.M. Wong, Lecture notes in physics, vol. 144 (Springer, Berlin, 1981) p. 248

57. G.F. Bertsch, Nucl. Phys. 74 (1965) 234

58. K. Andō, H. Bandō and S. Nagata, Suppl. of Prog. Theor. Phys. No. 65 (1979) 1, and references quoted therein

59. J. Shurpin, Ph.D. thesis (State University of New York at Stony Brook, 1980, unpublished);
J. Shurpin, T.T.S. Kuo and D. Strottman, Nucl. Phys. A354 (1981) 589C, and to be published

60. J.P. Vary, P.U. Sauer and C.W. Wong, Phys. Rev. C7 (1973) 1776

61. C.L. Kung, T.T.S. Kuo and K.F. Ratcliff, Phys. Rev. C19 (1979) 1063

62. M. Sommermann, H. Müther, K.C. Tam, T.T.S. Kuo and A. Faessler, Phys. Rev. C23 (1981) 1765

63. G. Höhler and E. Pietarinen, Nucl. Phys. B95 (1975) 210

64. K.C. Tam, H. Müther, M. Sommermann, T.T.S. Kuo and A. Faessler, Nucl. Phys. A361 (1981) 412

65. T.T.S. Kuo and E.M. Krenciglowa, Nucl. Phys. A342 (1980) 454

66. N. Austern, Direct nuclear reaction theory (Wiley, New York, 1970)

67. H. Feshbach, Ann. Rev. Nucl. Sci. 8 (1958) 49

68. J.S. Bell and E.J. Squires, Phys. Rev. Lett. 3 (1959) 96

69. J.T. Londergan and G.A. Miller, Phys. Rev. Lett. 46 (1981) 1545; Phys. Rev. C25 (1982) 46

70. S.Y. Lee, F. Osterfeld, K.C. Tam and T.T.S. Kuo, Phys. Rev. C24 (1981) 329

71. B. Buck and R. Lipperheide, Nucl. Phys. A368 (1981) 141

72. G. de Guzman, T.T.S. Kuo, K. Holinde, R. Machleidt, A. Faessler and H. Müther, Nucl. Phys. A443 (1985) 601

73. M.B. Johnson, Ann. Phys. (N.Y.) 97 (1976) 400

74. I. Lindgren and J. Morrison, Atomic many-body theory, 2nd ed. (Springer, Berlin, 1986)

75. E.U. Condon and H. Odabasi, Atomic structure (Cambridge Univ. Press, Cambridge, 1980)

76. R.D. Cowan, The theory of atomic structure and spectra (Univ. of California, Berkeley, 1981)

77. C. Froese-Fischer, The Hartree-Fock method for atoms (Wiley, New York, 1977)

78. J.D. Jackson, Classical electrodynamics, 2nd ed. (Wiley, New York, 1975)

79. A. deShalit and H. Feshbach, Theoretical nuclear physics, vol. 1: Nuclear structure (Wiley, New York, 1974)

80. I. Lindgren and S. Lundqvist (eds.), Many-body theory of atomic systems, Phys. Scripta 21, No. 3/4 (1980)

81. J. Morrison and S. Salomonson, Phys. Scripta 21 (1980) 343

82. P.G.H. Sandars, Adv. Chem. Phys. 14 (1969) 365

83. S. Salomonson, I. Lindgren and A.-M. Mårtensson, Phys. Scripta 21 (1980) 351

84. C.F. Bunge, Phys. Rev. A14 (1976) 1965

85. S. Bashkin and J.A. Stoner, Jr., Atomic energy levels and Grotrian diagrams (North-Holland, Amsterdam, 1975)

86. B.H. Wildenthal, Prog. in Part. and Nucl. Phys. 11 (1984) 5;
B.A. Brown and B.H. Wildenthal, Ann. Rev. Nucl. Part. Sci. 38 (1988) 29

87. M. Karplus and R.N. Porter, Atoms and molecules (Benjamin, New York, 1970)

88. I.N. Levine, Molecular spectroscopy (Wiley, New York, 1975)

89. U. Kaldor, Phys. Rev. Lett. 31 (1973) 1338

90. P.S. Stern and U. Kaldor, J. Chem. Phys. 64 (1976) 2002

91. G. Hose and U. Kaldor, J. Phys. B12 (1979) 3827, and references quoted therein;
U. Kaldor, in Condensed matter theories, ed. J. Keller, vol. 4 (Plenum, New York, 1989) p. 67

Lecture Notes in Mathematics

Vol. 1236: Stochastic Partial Differential Equations and Applications. Proceedings, 1985. Edited by G. Da Prato and L. Tubaro. V, 257 pages. 1987.

Vol. 1237: Rational Approximation and its Applications in Mathematics and Physics. Proceedings, 1985. Edited by J. Gilewicz, M. Pindor and W. Siemaszko. XII, 350 pages. 1987.

Vol. 1250: Stochastic Processes – Mathematics and Physics II. Proceedings 1985. Edited by S. Albeverio, Ph. Blanchard and L. Streit. VI, 359 pages. 1987.

Vol. 1251: Differential Geometric Methods in Mathematical Physics. Proceedings, 1985. Edited by P.L. García and A. Pérez-Rendón. VII, 300 pages. 1987.

Vol. 1255: Differential Geometry and Differential Equations. Proceedings, 1985. Edited by C. Gu, M. Berger and R.L. Bryant. XII, 243 pages. 1987.

Vol. 1256: Pseudo-Differential Operators. Proceedings, 1986. Edited by H.O. Cordes, B. Gramsch and H. Widom. X, 479 pages. 1987.

Vol. 1258: J. Weidmann, Spectral Theory of Ordinary Differential Operators. VI, 303 pages. 1987.

Vol. 1260: N.H. Pavel, Nonlinear Evolution Operators and Semigroups. VI, 285 pages. 1987.

Vol. 1263: V.L. Hansen (Ed.), Differential Geometry. Proceedings, 1985. XI, 288 pages. 1987.

Vol. 1265: W. Van Assche, Asymptotics for Orthogonal Polynomials. VI, 201 pages. 1987.

Vol. 1267: J. Lindenstrauss, V.D. Milman (Eds.), Geometrical Aspects of Functional Analysis. Seminar. VII, 212 pages. 1987.

Vol. 1269: M. Shiota, Nash Manifolds. VI, 223 pages. 1987.

Vol. 1270: C. Carasso, P.-A. Raviart, D. Serre (Eds.), Nonlinear Hyperbolic Problems. Proceedings, 1986. XV, 341 pages. 1987.

Vol. 1272: M.S. Livšic, L.L. Waksman, Commuting Nonselfadjoint Operators in Hilbert Space. III, 115 pages. 1987.

Vol. 1273: G.-M. Greuel, G. Trautmann (Eds.), Singularities, Representation of Algebras, and Vector Bundles. Proceedings, 1985. XIV, 383 pages. 1987.

Vol. 1275: C.A. Berenstein (Ed.), Complex Analysis I. Proceedings, 1985–86. XV, 331 pages. 1987.

Vol. 1276: C.A. Berenstein (Ed.), Complex Analysis II. Proceedings, 1985–86. IX, 320 pages. 1987.

Vol. 1277: C.A. Berenstein (Ed.), Complex Analysis III. Proceedings, 1985–86. X, 350 pages. 1987.

Vol. 1283: S. Mardešić, J. Segal (Eds.), Geometric Topology and Shape Theory. Proceedings, 1986. V, 261 pages. 1987.

Vol. 1285: I.W. Knowles, Y. Saitō (Eds.), Differential Equations and Mathematical Physics. Proceedings, 1986. XVI, 499 pages. 1987.

Vol. 1287: E.B. Saff (Ed.), Approximation Theory, Tampa. Proceedings, 1985–1986. V, 228 pages. 1987.

Vol. 1288: Yu. L. Rodin, Generalized Analytic Functions on Riemann Surfaces. V, 128 pages, 1987.

Vol. 1294: M. Queffélec, Substitution Dynamical Systems – Spectral Analysis. XIII, 240 pages. 1987.

Vol. 1299: S. Watanabe, Yu.V. Prokhorov (Eds.), Probability Theory and Mathematical Statistics. Proceedings, 1986. VIII, 589 pages. 1988.

Vol. 1300: G.B. Seligman, Constructions of Lie Algebras and their Modules. VI, 190 pages. 1988.

Vol. 1302: M. Cwikel, J. Peetre, Y. Sagher, H. Wallin (Eds.), Function Spaces and Applications. Proceedings, 1986. VI, 445 pages. 1988.

Vol. 1303: L. Accardi, W. von Waldenfels (Eds.), Quantum Probability and Applications III. Proceedings, 1987. VI, 373 pages. 1988.

Lecture Notes in Physics

Vol. 340: M. Peshkin, A. Tonomura, The Aharonov-Bohm Effect. VI, 152 pages. 1989.

Vol. 341: E.F. Milone (Ed.), Infrared Extinction and Standardization. Proceedings, 1988. III, 79 pages. 1989.

Vol. 342: M. Balabane, P. Lochak, C. Sulem (Eds.), Integrable Systems and Applications. Proceedings, 1988. VII, 342 pages. 1989.

Vol. 343: M. Month, S. Turner (Eds.), Frontiers of Particle Beams; Observation, Diagnosis and Correction. Proceedings, 1988. IX, 509 pages. 1989.

Vol. 344: M. Rascle, D. Serre, M. Slemrod (Eds.), PDEs and Continuum Models of Phase Transitions. Proceedings, 1988. VII, 229 pages. 1989.

Vol. 345: H. Holden, A. Jensen (Eds.), Schrödinger Operators. Proceedings, 1988. V, 458 pages. 1989.

Vol. 346: D. Lüst, S. Theisen, Lectures on String Theory. VII, 346 pages. 1989.

Vol. 347: Ph. Blanchard, J.-P. Dias, J. Stubbe (Eds.), New Methods and Results in Non-linear Field Equations. Proceedings. 1987. VII, 133 pages. 1989.

Vol. 348: A. Bohm, M. Gadella, Dirac Kets, Gamow Vectors and Gel'fand Triplets. VII, 119 pages. 1989.

Vol. 349: Y.A. Kubyshin, J.M. Mourão, G. Rudolph, I.P. Volobujev, Dimensional Reduction of Gauge Theories, Spontaneous Compactification and Model Building. X, 80 pages. 1989.

Vol. 350: G. Tenorio-Tagle, M. Moles, J. Melnick (Eds.), Structure and Dynamics of the Interstellar Medium. Proceedings, 1989. XXI, 537 pages. 1989.

Vol. 351: A. Dervieux, B. Larrouturou (Eds.), Numerical Combustion. Proceedings, 1989. VII, 481 pages. 1989.

Vol. 352: K.B. Wolf (Ed.), Lie Methods in Optics II. Proceedings, 1988. XII, 197 pages. 1989.

Vol. 353: M. Barthes, J. Léon (Eds.), Nonlinear Coherent Structures. Proceedings, 1989. X, 280 pages. 1990.

Vol. 354: J. De Coninck, F. Dunlop (Eds.), Wetting Phenomena. Proceedings, 1988. VII, 112 pages. 1990.

Vol. 355: R. Lima, L. Streit, R. Vilela Mendes (Eds.), Dynamics and Stochastic Processes. Proceedings, 1988. V, 298 pages. 1990.

Vol. 356: R. Wehrse (Ed.), Accuracy of Element Abundances from Stellar Atmospheres. Proceedings, 1988. V, 109 pages. 1990.

Vol. 357: J.L. Lumley (Ed.), Whither Turbulence? Turbulence at the Crossroads. Proceedings, 1989. V, 535 pages. 1990.

Vol. 358: J. Kovalevsky, Astrométrie moderne. VI, 294 pages. 1990.

Vol. 359: K. Kirchgässner (Ed.), Problems Involving Change of Type. Proceedings, 1988. XII, 207 pages. 1990.

Vol. 360: Y. Mellier, B. Fort, G. Soucail (Eds.), Gravitational Lensing. Proceedings 1989. XV, 315 pages. 1990.

Vol. 361: P. Gaigg, W. Kummer, M. Schweda (Eds.), Physical and Nonstandard Ganges. Proceedings, 1989. IX, 310 pages. 1990.

Vol. 362: N.E. Kassim, K.W. Weiler (Eds.), Low Frequency Astrophysics from Space. Proceedings. XII, 280 pages. 1990.

Vol. 363: V. Ruždjak, E. Tandberg-Hanssen (Eds.), Dynamics of Quiescent Prominences. Proceedings, 1989. XI, 304 pages. 1990.

Vol. 364: T.T.S. Kuo, E. Osnes, Folded-Diagram Theory of the Effective Interaction in Nuclei, Atoms and Molecules. V, 175 pages. 1990.

W. Glöckle, University of Bochum

The Quantum Mechanical Few-Body Problem

1983. VIII, 197 pp. 17 figs. (Texts and Monographs in Physics) Hardcover DM 104,–
ISBN 3-540-12587-6

This is a self-contained introduction to the quantum mechanical few-body problem requiring only a basic knowledge of standard quantum mechanics. The author presents modern formulations of one to four particle models. In the examples – spin-observables in nucleon-nucleon scattering and the three-nucleon bound state – he shows the way to their numerical treatment. In gradually building up the theory the author relies on the complete set of Lippmann-Schwinger equations which provide a unified and transparent approach.

As a textbook, the derivations of few-body euqations including the system of four particles, the discussion of their properties like asymptotic behavior and spuriosities, and the detailed descriptions of how to handle the equations in actual numerical calculations is new and will be of benefit to graduate students and research workers. This presentation fills a gap in available monographs on this subject.

P. Ring, P. Schuck, Technical University of Munich

The Nuclear Many-Body Problem

1980. XVII, 716 pp. 171 figs. (Texts and Monographs in Physics) Hardcover DM 102,–
ISBN 3-540-09820-8

This book, while covering a fair amount of physical observations, stresses the methodology and technical aspects of the different theories presently used in the desription of the nucleus. The authors present the more modern theories such as Boson expansions, generator coordinates, time-dependent Hartree-Fock method, and semiclassical models which so far have found only limited mention in textbooks. The book also covers subjects like the liquid drop and the shell model, both presented in an updated version in, for example, rotations and random phase approximation. The full presentation of mathematical details, illustrated by observational data, will help the student fully understand the present views on the nuclear many-body problem.